STATISTICS
Third Edition

Donald J. Koosis
ISC Consultants, Inc.
New York, New York

IN CONSULTATION WITH

Arthur P. Coladarci
School of Education
Stanford University

JOHN WILEY & SONS
New York • Chichester • Brisbane • Toronto • Singapore

To Irene with much love.

Publisher: Stephen Kippur
Editor: Elizabeth G. Perry
Managing Editor: Katherine Schowalter
Composition & Make-up: Ganis & Harris, Inc.

Library of Congress Cataloging in Publication Data

Koosis, Donald J.
  Statistics.

  Includes index.
  1. Statistics--Programmed instruction.
I. Coladarci, Arthur P.   II. Title.
QA276.12.K667   1985   519.5'07'7   85-6292

ISBN 0-471-82720-7

Printed in the United States of America

85 86 10 9 8

STATISTICS

More than 100 Wiley Self-Teaching Guides teach practical skills on everything from accounting to astronomy, from microcomputers to mathematics.

STGs FOR MATHEMATICS

Accounting Essentials, 2nd Edition, Margolis & Harmon
Finite Mathematics, Rothenberg
Geometry and Trigonometry for Calculus, Selby
Linear Algebra with Computer Applications, Rothenberg
Management Accounting, Madden
Math Shortcuts, Locke
Math Skills for the Sciences, Pearson
Practical Algebra, Selby
Quick Algebra Review, Selby
Quick Arithmetic, 2nd Edition, Carman & Carman
Quick Calculus, 2nd Edition, Kleppner & Ramsey
Statistics, 3rd Edition, Koosis

Look for these and other Self-Teaching Guides at your favorite bookstore!

To the Reader

This book is designed for the individual who wants to learn the basic concepts and procedures of statistics as it is used today for educational, professional, or personal reasons. You may be interested in statistics as part of a liberal education. You may be a student of or employed in the field of education, engineering, or one of the social, biological, or physical sciences. The methods taught here are common to all these fields. You will find examples of the application of statistical techniques to each of these fields in the book.

This is a "how to" book. It will teach you how to perform a number of useful statistical tests and procedures. It will teach you how to determine which of these tests and procedures are appropriate under given circumstances. Although you will learn the basic concepts on which the tests are based, you will not learn the mathematical derivation of every formula nor all the possible variations of every test. The book assumes only a knowledge of elementary algebra—enough to substitute numbers for symbols in a formula and to keep the + and − signs straight when you are working with positive and negative numbers.

Computers are generally used for statistical calculations when large amounts of data are analyzed and extensive, repetitive calculations are required. In this book, we provide examples of the output of computer programs for statistical analysis. However, we emphasize the kinds of statistical work that can be done "by hand"—problems with small amounts of data and relatively simple computations. Such problems can be completed with a pencil and paper or a simple electronic calculator. You will be able to use these techniques for yourself in everyday work and study. Since the computer techniques are an extension of the same basic methods taught here, this book will also help prepare you to become a more sophisticated consumer of these services.

As you work through this book, remember that the results of statistical tests and procedures are only as meaningful as the data on which they are based. The problems of measurement and experimental design are beyond the scope of this book and are touched on only in passing. In many cases you will have to apply knowledge of your field from other sources when you begin to put your statistical skills to use. At the end of the book is a test that you may use as a final examination to evaluate your progress. It includes material from all chapters. If you successfully complete the review problems for each chapter and the test, you may be confident that you have mastered the material in this book.

DONALD J. KOOSIS

New York, New York

How to Use This Book

This book is organized into numbered sections called frames. Each frame presents some new material, asks you to answer a question, and gives the correct answer. When you study the book, use an index card or note pad to cover the book's answers. Slide the card down the page until you come to a dotted line. Write your answer: then move the card down to check your answer against the correct one. Most of the time you should be able to answer correctly. If you find that you have made an error, look back at the preceding material to make sure you understand the correct answer before you go on.

This is not a book you can read through quickly. Statistics, like other mathematical subjects, requires concentration. Each chapter represents about two hours' work. If possible, try to complete a chapter in one or two sessions and stop only at the end of a section or chapter. Frequent interruptions make the material more difficult to master.

All the computations required in this book can be worked out by hand with reasonable effort; however, there is no special virtue in doing the arithmetic by hand. If possible, you should use an electronic calculator. The best choice is one that computes squares and square roots automatically. (It will have keys labeled x^2 and $\sqrt{x}$.) If you have a programmable calculator, work through the first few problems of each type step by step before you try to use the programming features.

If your interest is primarily in understanding the steps involved in a test, and you have only limited patience for arithmetic, you may choose to skip some of the computations. When you make this choice, be sure nevertheless that you set up the problem and compare your work, as far as it goes, with the solution in the answer.

At the end of each chapter is a set of review problems. These problems not only review the content of the chapter; they also relate it to other material you have learned. Working these problems is an important part of the study process. Do not skip them.

At the end of the book is a test that you may use as a final examination to evaluate your progress.

Contents

STATISTICS

CHAPTER ONE
Basic Skills

This chapter is about how to summarize data. When you perform an experiment, conduct a survey, or collect information, you usually wind up with a number of observations. For example, you have the test scores of 27 classmates, or the opinions of 12 jurors, or the height of 100 redwood trees. Your first problem is to summarize this information in some way that lets you generalize—that lets you see the forest as well as the individual trees.

One common and useful way to summarize a group of observations is to draw a graph—a *frequency-distribution graph*. Another way is to compute some sort of "average" that describes a typical observation—a measure of central tendency. Three commonly used measures of central tendency are the *mean*, the *median*, and the *mode*.

It is also useful to describe numerically how much the observations differ from one another—what we call their *variability*. Three common measures of variability are the *range*, the *standard deviation*, and *variance*.

This chapter will teach you the most common methods of summarizing data. When you have completed this chapter you will be able to:

● construct a frequency-distribution graph;
● recognize and apply some common vocabulary to describe frequency distributions;
● compute three measures of central tendency: the mean, the median, and the mode;
● compute three measures of variability: the range, the standard deviation, and the variance.

HOW TO CONSTRUCT A FREQUENCY-DISTRIBUTION GRAPH

The frequency distribution is a useful summary of most kinds of data. A frequency distribution sorts observations into categories and describes how often observations fall into each category (either as a number or a percent). Very likely you have used frequency distributions to analyze the data of your own daily life.

1. Consider the following data:

    Ann has green eyes   Alice has blue eyes
    Jane has brown eyes   Judy has blue eyes
    Marie has brown eyes   Sue has brown eyes
    Joan has blue eyes   Carol has brown eyes
    Victoria has brown eyes   Barbara has brown eyes

    How would you summarize these data?

    –

    You probably used a table like this:

    brown eyes   6 or 60%
    blue eyes   3 or 30%
    green eyes   1 or 10%

    You might also have drawn a frequency-distribution bar graph like the one below.

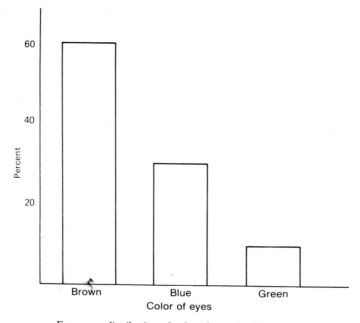

Frequency distribution of color of eyes for 10 women.

2. A bar graph like the one opposite is a useful summary of the data in question. This bar graph shows the _____ of eye colors.

_ _ _ _ _ _ _ _ _ _ _ _ _ _ _ _ _ _ _ _

frequency distribution

3. If you used a computer to analyze your data, you could obtain a printout that looks something like this:

VALUE LABEL	FREQUENCY	PERCENT
BROWN	6	60.0
BLUE	3	30.0
GREEN	1	10.0

Your printout might also be in graphic form, like this:

COUNT	VALUE LABEL	ONE SYMBOL EQUALS APPROX 1.00 OCCURRENCES
6	BROWN	* * * * * *
3	BLUE	* * *
1	GREEN	*

Do these printouts contain any additional information not in the table and bar graph of frames 1 and 2? (Yes/No)

_ _

No

4. Suppose you are interested in the types of cars customers of a particular shopping center drive. You observe the following types of car in the parking lot on a given afternoon:

4-door sedan	compact	compact
station wagon	4-door sedan	hardtop
hardtop	4-door sedan	station wagon
sports car	compact	station wagon
4-door sedan	4-door sedan	pick-up truck
sports car	station wagon	hardtop
station wagon	station wagon	hardtop
sports car	4-door sedan	compact
compact	compact	hardtop
convertible	compact	station wagon
antique car	4-door sedan	4-door sedan
station wagon	4-door sedan	4-door sedan
station wagon	station wagon	

Prepare a frequency-distribution bar graph of these observations.

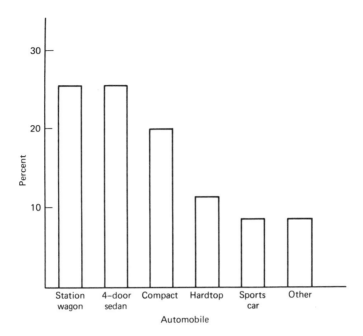

Your graph should look something like the one above. As is customary, it measures the number of observations as a percentage along the vertical axis of the graph and shows the categories along the horizontal axis. The order in which you list the categories is not critical, although it helps the reader if there is some logic to the order; for example, here we have listed the categories from highest to lowest frequency.

5. Often, instead of categories, the data we wish to summarize are measurements on a continuous scale; for example, length or time or temperature. Then we must group the measurements into categories; for example, look at the following raw data and the frequency distribution based on these temperatures:

60.0°	62.0°	70.0°	72.5°	74.5°
60.0	62.0	70.5	72.5	74.5
60.5	63.5	71.0	73.0	74.5
61.0	63.5	71.0	73.0	75.0
61.0	64.0	71.5	73.0	75.0
61.0	67.5	72.0	73.0	75.0
61.0	68.0	72.0	73.5	76.0
62.0	69.5	72.0	73.5	76.5
62.0	70.0	72.5	73.5	77.0
62.0	70.0	72.5	74.0	79.0

Temperature in Degrees

The first bar in the graph above represents measurements between _____ and _____ °.

– – – – – – – – – – – – – – – – – – – –

59.25 and 61.25°

6. Are there any measurements that fall exactly on the dividing line between two categories?

_ _

No

7. Why do you suppose no measurements fall on the dividing line between two categories?

(a) It was just luck.
(b) We chose the categories with that in mind.

_ _

(b) We chose the categories with that in mind.

8. How many categories are used to group the data?

_ _

10

9. The following illustrations (Graphs A, B, C) show three frequency-distribution graphs based on the same data. One is the original graph which uses 10 categories, another uses 40 categories, and the third uses only three categories. Which one do you think presents the most useful summary of the data?

Graph A

Graph B

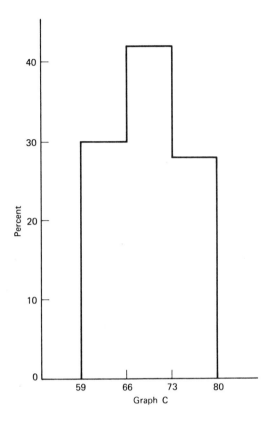

Graph C

- - - - - - - - - - - - - - - - - - - -

For most purposes Graph A probably presents the most useful summary. Graph B is so detailed that the over-all pattern is not clear, whereas Graph C goes to the other extreme and obscures a difference that is large enough to be important.

10. As a rule of thumb, it is usually best to choose categories so that there are between 10 and 20 categories in a distribution of measurements; for example, if you are preparing a distribution of test scores that range from 50 to 100 points, which set of categories would be preferred?

(a) 50–52, 53–54, 55–56, etc.?
(b) 50–59, 60–69, 70–79, etc.?
(c) 50–54, 55–59, 60–64, etc.?

_ _

(c) 50–54, 55–59, 60–64, etc.

11. To set the exact limits of categories consider the accuracy of your measurements; for example, the measurements in frame 4 were taken with a thermometer that was accurate only to the nearest half degree. By making the exact limits of the first category 59.25 and 61.25 we could be sure that no measurements would fall on the dividing line between two categories.

If you are preparing a distribution of weights that are rounded to the nearest gram, which would be the better dividing line between two categories?

(a) 5 gm?
(b) 5.5 gm?

_ _

(b) 5.5 gm

12. Using 10 categories, construct a frequency-distribution graph for the following data:

1.1	1.8	2.1	2.2	2.7	3.6
1.4	1.8	2.1	2.3	2.9	3.8
1.4	1.9	2.1	2.4	2.9	4.1
1.4	1.9	2.1	2.4	3.0	4.5
1.7	1.9	2.1	2.5	3.2	4.9

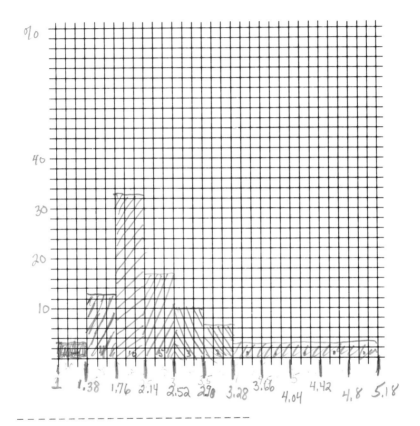

Your graph should look something like this one. We used an exact limit for the first category of 1.05–1.55, but you might equally well have made the limits 0.95–1.45.

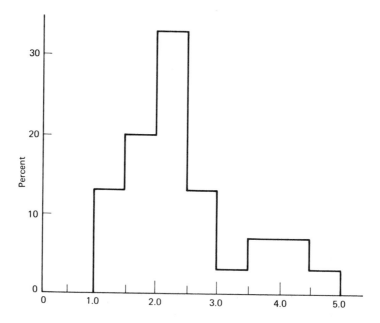

DESCRIBING FREQUENCY DISTRIBUTIONS

There are some technical terms commonly used to describe frequency distributions. You will find it useful to know these terms: mode, bimodal, skewed, normal.

13. Distributions of measurements can have various shapes. Some of the more common situations are illustrated below. Assume that each of these distributions represents the times required by a group of 100 college students to solve a given Chinese puzzle. The time limit for each puzzle was half an hour.

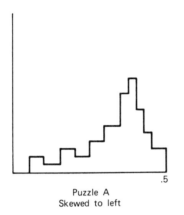

Puzzle A
Skewed to left

Puzzle B
Approximately normal

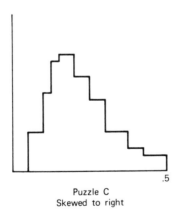

Puzzle C
Skewed to right

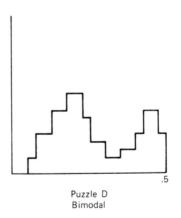

Puzzle D
Bimodal

For one of these puzzles there were two obvious approaches to solving it. About half the students started with the correct approach and solved it quite quickly, whereas the other half attempted the wrong approach first and took substantially longer. Which puzzle meets this description and what name is applied to the distribution?

_ _

Puzzle D
Bimodal

14. For one of the puzzles there was one clear average time. About as many finished faster than average as finished slower than average and the distribution was bell shaped. Which puzzle fits this description and what name is applied to the distribution?

_ _

Puzzle B
Approximately normal

The normal distribution has a precise mathematical definition and you will learn more about it later.

15. Just working with one of the puzzles for a fixed amount of time automatically resulted in a solution. Most students simply kept working on the puzzle until it solved itself at the end of this time, but a moderate number were able to come up with faster solutions. Which puzzle fits this description and what name is applied to its distribution?

_ _

Puzzle A
Skewed to the left

16. The peak, or mode, of a distribution is the most common score; that is, the tallest bar in the graph. What is the mode of this set of data?

3, 3, 4, 4, 4, 5, 5, 5, 5, 5, 5, 5, 7, 8, 8, 9, 9

_ _

The mode is 5, the most common score.

17.

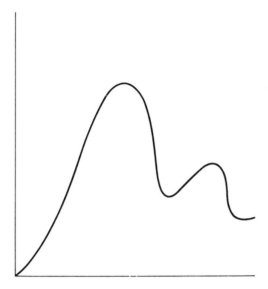

A peak in a <u>distribution</u>, as shown above, is called a mode. When a distribution has two peaks, it is called _____.

- - - - - - - - - - - - - - - - - - -

bimodal

18. Look at the distributions below:

Skewed to left

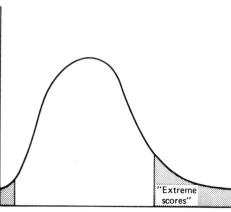

Skewed to right

When there is a greater number of extreme scores on the left side of a distribution than on the right side, we may say it is skewed to the (left/right).

_ _ _ _ _ _ _ _ _ _ _ _ _ _ _ _ _ _ _ _

left

19. Sketch a distribution that is skewed to the right.

20. Sketch an approximately normal distribution.

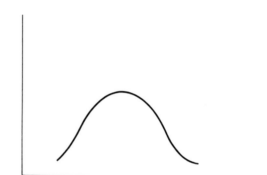

21. Sketch a bimodal distribution.

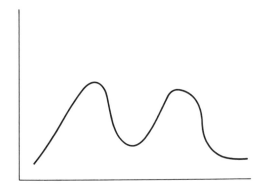

22. What is the mode of this distribution?

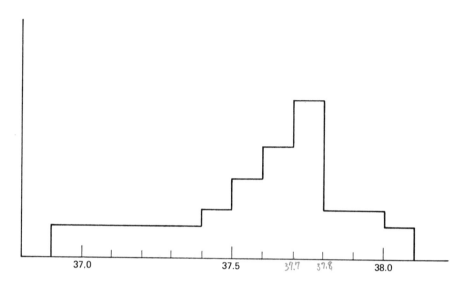

- - - - - - - - - - -

37.7 – 37.8

23. What is the mode of this distribution?

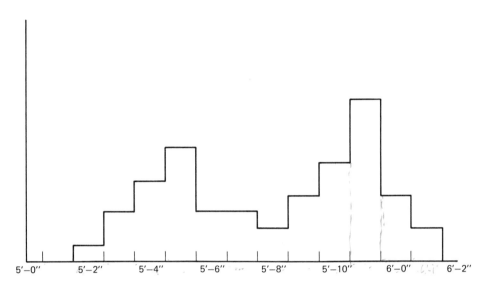

----

5 ft 5 in
and
5 ft 11 in.

24.   Describe each of these distributions in a few words.

(a)

(b)

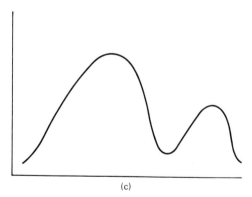

(c)

— — — — — — — — — —

(a)   Approximately normal
(b)   Skewed to the left
(c)   Bimodal

MEASURES OF CENTRAL TENDENCY

Often, we want to summarize a whole distribution of measurements with a single number that is typical of the distribution. This kind of number is called a "measure of central tendency." There are three measures of central tendency you should know about — they are the mean, the median, and the mode. Since you already know about the mode, you have only the median and the mean left to learn.

25. | The median is the middle measurement in a set of measurements. |

To find the median you must
(a)   sort the observations in order of magnitude
(b)   then find the middle number
For example, in the following set of 11 measurements, the sixth is the median:

11, 11, 13, 15, 17, 17, 17, 19, 19, 19, 19

The median of this distribution is _____.

_ _

17

26.   What is the median of the following set of measurements?

2.0, 5.8, 6.1, 10.5, 53.9, 54.0, 78.6

_ _

10.5

27.   What is the median of the following set of observations? What is the mode?

55 months, 50, 53, 54, 53, 54, 55, 56, 58, 60, 54, 54, 55, 58, 52, 54, 56, 57, 59

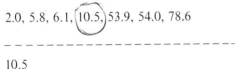

Median, 55
Mode, 54

When there is an even number of observations, the middle of the list is halfway between two observations. The usual procedure is to call the halfway point the median; for example,

1, 5, 6, 9, 11, 12

↑
median = 7.5

28.

> The mean is the value obtained by adding all the measurements and dividing by the number of measurements.

For example,

```
 1.0
 2.0
 3.0
 6.0
 6.0
 7.0
 9.0
─────
34.0
```

$$\overset{4.9}{7)\ \overline{34.0}} \longleftarrow \text{mean}$$

For this series of observations, find the mean, median, and mode. (If you have an electronic calculator, use it.)

400, 600, 800, 800, 850, 850, 850, 850, 850, 950, 1000

‒ ‒

Mean, 800 (8800/11)
Median, 850
Mode, 850

29. Let's look at the formula for the mean. It looks a little alarming at first, but since you have already computed a mean you know that the process is simple.

> The formula for the mean is
>
> $$\mu = \frac{\Sigma x}{n}$$
>
> In this formula
>
> μ (Greek letter "mu") stands for the mean,
> x stands for each of the individual observations,
> Σ (summation sign) indicates the operation of summing all the values of x,
> n stands for the number of observations.

We can read the formula "mu equals summation x over n." Copy the formula once to get used to it.

‒ ‒

$$\mu = \frac{\Sigma x}{n}$$

30. Now let us apply the formula to some numbers. The following are the durations in minutes of a series of telephone calls. You want to know the mean duration of the calls.

1.0, 5.0, 5.3, 6.0, 6.0, 7.0, 7.7, 8.0, 8.1, 8.9

$$\mu = \frac{\Sigma x}{n}$$

In this problem x has a series of different values. What are they?

_ _

1.0, 5.0, 5.3, etc., the duration of each call

31. Σx is the sum of all the values of x. What is Σx in this case?

_ _

63.0 (1.0 + 5.0 + 5.3 + ..., etc.)

32. What is n in this case?

_ _

10 (the number of observations you have for x)

33. What is μ?

_ _

$$\mu = \frac{\Sigma x}{n} = \frac{63.0}{10} = 6.3$$

MEASURES OF VARIABILITY

A measure of variability is a way of indicating how dispersed a set of observations is. The range of a distribution—the difference between the largest observation in the distribution and the smallest—is one crude measure of variability. The other measures of variability you will learn about in this book are the standard deviation and the variance.

Even though a computer will often compute standard deviations and variances for you, we strongly suggest you use a calculator or pencil and paper to complete the computations in this section. Working through the process will help you remember the meaning of the formulas.

34. Standard deviation and variance are measures of _____.

- -

variability

35.
> The basic formula for standard deviation is
>
> $$\sigma = \sqrt{\frac{\Sigma(x-\mu)^2}{n}}$$

σ (Greek letter "sigma") stands for the standard deviation. The other symbols should be familiar to you. If your math is a little rusty, don't panic! We will walk through the formulas step by step. (If you have already seen a somewhat different formula be patient; we will get to it in a moment.)

Let us apply the formula step-by-step to the following data:

Step 1. Find the mean, μ (see shaded area 1).

Step 2. Find $(x-\mu)$ for each of the different values of x. The parentheses indicate that this step comes first (see shaded area 2).

Step 3. Find $(x-\mu)^2$ for each of the different values of x and then find the sum of all these values of $(x-\mu)^2$ (see shaded area 3).

Step 4. Divide $\Sigma(x-\mu)^2$ by the number of different values of x (see shaded area 4).

Step 5. Find the square root of $[\Sigma(x-\mu)^2]/n$. (Use a calculator or look it up in a table like Table II in the appendix; see shaded area 5.)

1.

x
2
2
2
3
3
4
4
4
$\overline{\Sigma x = 24}$

$$\mu = \frac{\Sigma x}{n} = \frac{24}{8} = 3$$

2. $(x - \mu)$

-1
-1
-1
0
0
$+1$
$+1$
$+1$

3. $(x - \mu)^2$

$+1$
$+1$
$+1$
0
0
$+1$
$+1$
$+1$
$\overline{\Sigma(x-\mu)^2 = 6}$

4. $\dfrac{\Sigma(x-\mu)^2}{n} = \dfrac{6}{8} = 0.75$

5. $\sigma = \sqrt{\dfrac{\Sigma(x-\mu)^2}{n}} = \sqrt{0.75} = 0.866$

Now apply the same procedure to the following new data by completing the steps below:

1.

x
1
2
2
2
3
4
4
4
5
$\overline{\Sigma x = 27}$

$$\mu = \frac{\Sigma x}{n} = 3$$

2. $(x - \mu)$

-2
-1
-1
-1
0
1
1
1
2

3. $(x - \mu)^2$

$\overline{\Sigma(x-\mu)^2 =}$
14

4. $\dfrac{\Sigma(x-\mu)^2}{n} = 1.5\overline{5}$

5. $\sigma = \sqrt{\dfrac{\Sigma(x-\mu)^2}{n}} = 1.247$

1.	x	2. $(x-\mu)$	3. $(x-\mu)^2$
	1	-2	$+4$
	2	-1	$+1$
	2	-1	$+1$
	2	-1	$+1$
	3	0	0
	4	$+1$	$+1$
	4	$+1$	$+1$
	4	$+1$	$+1$
	5	$+2$	$+4$

$x = 27.$

$$\mu = \frac{\Sigma x}{n} = \frac{27}{9} = 3$$

$$\Sigma(x-\mu)^2 = 14$$

4. $$\frac{\Sigma(x-\mu)^2}{n} = \frac{14}{9} = 1.556$$

5. $$\sigma = \sqrt{\frac{\Sigma(x-\mu)^2}{n}} = \sqrt{1.556} = 1.248$$

36. The reasons for using this particular formula as the usual measure of variability depend on mathematical considerations that are beyond the scope of this book. Other formulas are possible, but this one happens to be the most useful. To make sure you understand the formula for standard deviation, let us restate it in other terms. First, the expression $(x - \mu)$ is simply the difference between each observation and the _____.

----

mean

37. Once the differences from the mean are computed, the next step is to

_____.

----

square them

38. The squared differences from the mean are then added up and divided by the number of observations; that is, they are _____.

_____

averaged

39. The final step is to _____.

_____

take the square root

40. "Standard deviation is the square root of the mean squared deviation from the mean." True or false?

_____

True

41. Compute the standard deviation of this set of numbers:

1, 1, 3, 5, 5 $\mu = 3$; $\Sigma(x-\mu)^2 = 16$ $\sigma = 1.789$

$x - \mu$

-2

-2

0

2

2

- -

x	$(x - \mu)$	$(x - \mu)^2$
1	$- 2$	$+ 4$
1	$- 2$	$+ 4$
3	0	0
5	$+ 2$	$+ 4$
5	$+ 2$	$+ 4$
$x = 15$		$\Sigma(x - \mu)^2 = 16$

$$\mu = \frac{\Sigma x}{n} = 3$$

$$\frac{\Sigma(x - \mu)^2}{n} = \frac{16}{5} = 3.2$$

$$\sigma = \sqrt{\frac{\Sigma(x - \mu)^2}{n}} = \sqrt{3.2} = 1.79$$

42. Variance is simply σ^2. If the standard deviation of a distribution is 2, what is its variance?

- -

4

43. If the variance of a distribution is 25.00, what is its standard deviation?

--- --- --- --- --- --- --- --- --- --- --- --- ---

5.00

44. You compute variance the same way that you compute the standard deviation, except that you skip the last step—taking the square root. So the formula for σ^2 is _____.

--- --- --- --- --- --- --- --- --- --- --- --- ---

$$\sigma^2 = \frac{\Sigma(x - \mu)^2}{n}$$

45. There is another formula for σ and σ^2 you will often see. It is mathematically equivalent to the one you have already learned, but it is easier to use for computation, since it avoids computing $(x - \mu)$ for each observation.

$$\sigma = \sqrt{\frac{\Sigma x^2 - (\Sigma x)^2/n}{n}}, \qquad \sigma^2 = \frac{\Sigma x^2 - (\Sigma x)^2/n}{n}$$

Let us work through this formula once to give you more practice in interpreting statistical formulas. It is not necessary to memorize the formula. The observations are:

1, 2, 2, 2, 3, 4, 4, 4, 5. $\mu = 3$ $\sigma^2 = 1.\overline{5}$

Read the descriptions of the steps and complete the table to find σ^2.

Step 1.   Find Σx by adding all the values of x. Then compute $(\Sigma x)^2$ by squaring Σx and divide by the number of values of x to find $(\Sigma x)^2/n$. Remember, perform the steps in parentheses first.

Step 2.   Find x^2 for each value of x and find the sum of all these values of x^2, Σx^2.

Step 3.   Compute $\Sigma x^2 - (\Sigma x)^2/n$ and then divide the result by n.

Complete the following computation:

1.

x
1
2
2
2
3
4
4
4
5

$\Sigma x =$

$(\Sigma x)^2 =$

$\dfrac{(\Sigma x)^2}{n} =$

2.

x^2
1
4
4
4
9
16
16
16
25

$\Sigma x^2 =$

3. $\sigma^2 = \dfrac{\Sigma x^2 - (\Sigma x)^2 / n}{n} =$

- -

1.

x
1
2
2
2
3
4
4
4
5

$\Sigma x = 27$

$(\Sigma x)^2 = 729$

$\dfrac{(\Sigma x)^2}{n} = \dfrac{729}{9} = 81$

2.

x^2
1
4
4
4
9
16
16
16
25

$\Sigma x^2 = 95$

3. $\sigma^2 = \dfrac{\Sigma x^2 - (\Sigma x)^2 / n}{n} = \dfrac{95 - 81}{9} = \dfrac{14}{9} = 1.56$

46. It is important to keep the steps in the right order when using statistical formulas.

$(\Sigma x)^2$ means "find the sum of the values of x, then square it."

Σx^2 means "find the values of _____, then find _____.

- -

x^2
the sum

47. $(\Sigma x)^2$ means

"First _____ then _____."

- -

find the sum
find the square

48. Σx^2 means
"First _____ then _____."

- -

find the square
find the sum

49. What does the symbol μ stand for?

- -

Mean

50. What does the symbol σ^2 stand for?

- -

Variance

51. What does the symbol σ stand for?

- -

Standard deviation

52. For the following data find μ, σ, σ^2, and the median. $\Sigma x = 21$ $\sigma = 2.07$
    Data: 1.0, 1.0, 2.0, 2.0, 3.0, 5.0, 7.0 $\mu = 3$ $\sigma^2 = 4.29$

- -

The median is the fourth observation in order of magnitude—2.0.

x	$(x - \mu)$	$(x - \mu)^2$
1	− 2	+ 4
1	− 2	+ 4
2	− 1	+ 1
2	− 1	+ 1
3	0	0
5	+ 2	+ 4
7	+ 4	+ 16
$\Sigma x = 21$		$\Sigma(x - \mu)^2 = 30$

$$\mu = \frac{\Sigma x}{n} = 3$$

$$\sigma^2 = \frac{\Sigma(x - \mu)^2}{n} = \frac{30}{7} = 4.29$$

$$\sigma = \sqrt{\frac{\Sigma(x - \mu)^2}{n}} = \sqrt{4.29} = 2.07$$

53. Read this computer output and fill in the appropriate information below.

VARIABLE: AGE

MEAN 37.16   MEDIAN 34.00   STD DEV 11.34

VARIANCE 128.60

VALID CASES 272

$\sigma =$ __11.34__ $\sigma^2 =$ __128.6__ $\mu =$ __37.16__
$n =$ __272__

- -

$\sigma = 11.34$ $\sigma^2 = 128.60$ $\mu = 37.16$
$n = 272$

REVIEW PROBLEMS

If you have successfully completed this chapter, you can now summarize data in a number of ways. You can:

- construct a frequency distribution;
- describe a frequency distribution as approximately normal, skewed or bimodal;
- compute the mean, median, and mode of your data;
- compute the standard deviation and variance of your data.

Since you can perform these computations yourself, you also know what the computer has done when it presents you with a report of these numbers.

Now try these review problems. Table I on pp. 253–256 lists any formulas you may need for reference. Table I is perforated so you can tear it out easily, if you wish.

1. Construct a frequency distribution for the following data:

0.1	2.5	2.6	5.1	5.3	6.7	7.1
7.3	7.5	7.5	8.9	9.9	10.1	11.3
11.7	12.5	12.8	14.1	15.0	17.5	18.9
19.8	21.7	24.4	24.9	30.2	34.3	37.3
38.9	44.2	47.3	53.0

(handwritten): $n = 25$
$\sum = 285.2$
$\mu \doteq 11.408$

2. Sketch a distribution that is skewed to the right.

3. What is the median of the data in question 1?

    10.1

4. What is the mean of the data in question 1?

    11.408

5. What is the standard deviation of the following data?
    3, 8, 8, 8, 9, 9, 9, 18 $\mu = 9$ $x - \mu = \begin{pmatrix} -6 \\ -1 \\ -1 \\ -1 \\ 0 \\ 0 \\ 0 \\ 9 \end{pmatrix}$ $(x-\mu)^2 = \begin{Bmatrix} 36 \\ 1 \\ 1 \\ 1 \\ \\ \\ \\ 81 \end{Bmatrix}$

    $\sigma^2 = 15$

    $\boxed{\sigma = \sqrt{15}}$

Answers

To review a problem, study the frames indicated after the answer.

1. The intervals used for the following distribution are 0.05 to 2.55, 2.55 to 5.05, and so on. You may have used different intervals, but your graph should be similar to this one. *See frames 1 to 12.*

2.

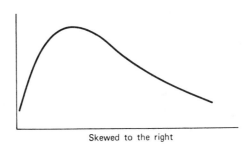

Skewed to the right

See frames 13 to 24.

3. Median = 10.1
    See frames 25 to 27.

4. $\mu = \dfrac{\Sigma x}{n} = \dfrac{285.2}{25} = 11.41$

    See frames 28 to 33.

5. $\sigma = \sqrt{\dfrac{\Sigma(x - \mu)^2}{n}} = \sqrt{\dfrac{120}{8}} = \sqrt{15} = 3.87$

    See frames 34 to 41.

CHAPTER TWO
Populations and Samples

The distinction between a *population* and a *sample* is very important in statistics. A *population* includes all possible observations of a particular type. For example, if you measured the height of every tree in the forest, you would be measuring a population. A *sample* includes only some of the observations, but selected in a way that gives every possible observation an equal chance of being chosen. For example, if you randomly chose 50 trees from a forest and measured their heights, you would be measuring a sample from a population.

Most of the techniques you will learn in this book have to do with using information about a sample to draw conclusions about the population, or vice versa. From information about a few randomly selected trees, you can draw some conclusions about the forest. Also, from information about the forest, you can draw some conclusions about a randomly selected group of trees.

The tool you use to go back and forth between samples and population is a mathematical table called a *sampling distribution*. In this chapter, you will get a brief taste of how sampling distributions are mathematically derived. You will also learn how to use two of the most useful sampling distribution tables to draw conclusions about a sample on the basis of information about the population.

When you have completed this chapter, you will be able to:

- distinguish between a population and a sample,
- use a sampling distribution,
- predict a sample proportion, using a binomial probability table.
- predict a sample mean using a normal probability table.

POPULATIONS AND SAMPLES

A population is all conceivable observations of a particular type. A sample is a limited number of observations from a population, chosen in a way that allows every possible observation an equal chance of occurring. Using statistics, it is possible to make statements about what a population is probably like on the basis of information from a sample. It is also possible to make statements about what samples will probably be like on the basis of information about the population.

1. In some cases, we use statistical techniques to go from information about a sample to information about a population. In other cases, we do just the

opposite and use information about a population to draw conclusions about the probable characteristics of a sample. For example, a researcher is interested in the learning ability of rats that receive a particular diet (she wants to compare the effect of this diet with that of another diet). There is no clear theoretical limit to the number of rats she could raise on the special diet, but for practical reasons she settles for measuring the learning ability of 50 rats. In this example the 50 rats are the _____.

_ _ _ _ _ _ _ _ _ _ _ _ _ _ _ _ _ _ _ _

sample

2. All rats who receive the special diet are the _____.

_ _

population

3. The researcher will

   (a) use information about the sample to make statements about what the population is probably like,

   (b) use information about the population to make statements about what the sample is probably like.

_ _

(a)

4. On the basis of certain physical laws, an engineer has determined that one out of every 27 components of a given type will be defective. He wants to determine the probability of finding two or more defective components in a particular batch of five.

   In this case the population is _____.

_ _

all components

5. The sample is _____.

_ _

the batch of five

6. The engineer will

   (a) use information about the sample to make statements about what the population is probably like,

   (b) use information about the population to make statements about what the sample is probably like.

   –

   (b)

7. Assume that the United States Census asks every inhabitant of the United States for his age and asks every 100th inhabitant for additional information about his education. The frequency distribution of ages obtained from these observations would be a (sample/population) distribution.

   –

   population (assuming no one was missed by the Census)

8. The frequency distribution of education levels would be a (sample/population) distribution.

   –

   sample

9. In order to find out which of the four supermarkets in your neighborhood has the best prices, you compile a typical shopping list and price the items on the list at all four stores. To decide whether the figures you obtain are samples or populations ask yourself, "Are the observations telling the complete story or am I just assuming that other observations will be similar?" Are these shopping list figures populations or samples?

   –

   Samples. The populations are all prices at each store; you are assuming that on other days and with other shopping lists you will obtain similar results.

10. A teacher wants to know the ages of the children in his class. He looks up the age of each child in the school records. The information he obtains is a (sample/population).

    –

    population

11. A number used to summarize a population distribution is called a parameter. A similar number used to describe a sample distribution is called a statistic; for example, if you are studying the population of the United States, the mean age of all inhabitants of the United States is a (parameter/statistic).

_ _

parameter

12. A researcher wants to estimate the average number of ladybugs per acre in Nebraska cornfields. To do so he counts the number of ladybugs in a large number of randomly selected one-acre plots. The mean number of ladybugs per acre in his sample is a (parameter/statistic).

_ _

statistic

13. The mean number of ladybugs per acre in Nebraska cornfields is a (parameter/ statistic).

_ _

parameter. It describes the state of affairs in the population of all Nebraska cornfields, though, of course, we have no way of counting every ladybug in every acre.

14. The mean of a sample distribution is a _____.

_ _

statistic

15. The mean of a population distribution is a _____.

_ _

parameter

16. The standard deviation of a population distribution is a _____.

– –

parameter

17. The standard deviation of a sample distribution is a _____.

– –

statistic

As we noted earlier, a sample should be chosen in such a way that every possible observation has an equal chance of occurring. However, this requirement is often very difficult to meet. Researchers in various fields have developed specialized techniques to assure random selection of samples. As a novice, you will probably be on safe ground if you learn and use the accepted methods of sampling for your field. But you should be aware that the proper selection of samples is one of the most difficult areas in the application of statistics. As you work through this book, you will find notes pointing out some of the difficulties. For a deeper understanding of these issues, you will need to study experimental design.

SAMPLING DISTRIBUTIONS

The idea of a sampling distribution is the key to our ability to reason back and forth from populations to samples and from samples to populations.

18. Let us develop a simple example of a sampling distribution. Take a coin out. You are interested in the proportion of times that the coin will come up heads when you flip it. We will call the event we are interested in a "success." Flip the coin twice and note how often it comes up heads. Write your results here:

first toss _____

second toss _____

The results of your two tosses are a (population/sample).

– –

sample

19. What is the population?

– –

The result of all possible tosses. Note that in this case the population is unlimited.

20. Look back at your sample. What proportion (p) of the time did the coin come up heads?

$$p = \frac{\text{number of successes (heads)}}{\text{number of tosses}} = \underline{\hspace{3cm}}.$$

- -

If your coin came up heads twice, $p = 1.0$
If your coin came up heads once, $p = 0.5$ (or $\frac{1}{2}$)
If your coin came up tails both times, $p = 0.0$

Note that p must always be between 1.0 and 0.0.

21. The value for p you just computed is a (parameter/statistic).

- -

statistic. It describes a sample. To distinguish between the parameter and the statistic, we use the capital letter P to identify the proportion of an entire population and the lowercase p to identify the proportion of a sample.

22. Think about your general experience with flipping coins. If you could perform a similar computation for the population, what value for P would you obtain?

$P = \underline{\hspace{1.5cm}}.$

- -

$P = 0.5$. Assuming that you are not carrying trick coins, the coin should come up heads half the time and tails half the time.

23. Take another sample of two tosses and compute p.

$p = \underline{\hspace{1.5cm}}.$

- -

If your coin came up heads twice, $p = 1.0$
If it came up heads once, $p = 0.5$
If it came up tails both times, $p = 0.0$

24. Take three more samples. Remember a sample consists of two tosses. You have now taken five samples of two tosses each. Summarize the results of your sampling below:

Sample   *p*

       1

       2

       3

       4

       5

Will the statistic *p* always equal 0.5?

– –

No. In fact it is unlikely that you found a value of 0.5 for all five samples.

25. If you continue to take a large number of samples of two and you average the values of *p* that you obtain from each sample, what do you think will be the average value of *p*?

– –

The average value of *p* will be approximately 0.5.

26. Can you construct a frequency distribution for the values of the statistic *p* that you obtain from a number of samples?

– –

Yes

27. Draw a frequency distribution graph for the five values of p you have obtained from different samples.

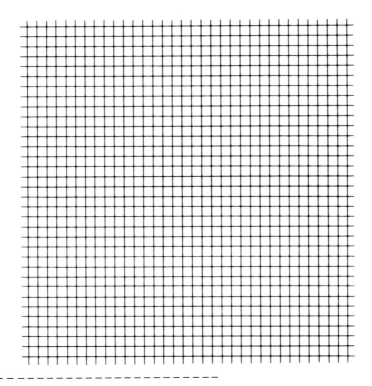

Your frequency distribution should look something like the following graphs. The mode will probably be 0.5, but not necessarily. Your graph could, for example, look like any of the graphs below.

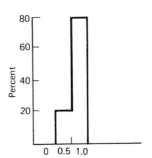

28. We can deduce mathematically what the frequency distribution of p would look like if you continued to take samples of two indefinitely. We do this by listing all possible results of two tosses and determining how probable each result is. First consider the first toss. What two results are possible?

_____

Heads or tails

29. Is either possible result more likely than the other?

_____

No

30. Now consider the possible results of the second toss. What are they?

_____

Again, either heads or tails

31. Will the results of the first toss have any influence on the second toss?

_____

No

32. Is either possible result of the second toss more likely than the other?

_____

No

33. We can summarize the possible results of the two tosses in the following table:

First toss   Second toss

Heads	Heads
	Tails
Tails	Heads
	Tails

As you can see, there are four possible results of the two tosses:

(a) Heads on the first toss and heads on the second toss
(b)
(c)
(d)

_ _

(b) Heads on the first toss and tails on the second toss
(c) Tails on the first toss and heads on the second toss
(d) Tails on the first toss and tails on the second toss

34. Is any of these results more likely than any of the others?

_ _

No, each of the four results is equally likely.

35. The possible results of the two tosses are

(a) heads   heads
(b) heads   tails
(c) tails   heads
(d) tails   tails

Next to each result write the corresponding value of the statistic p.

$$p = \frac{\text{number of successes}}{\text{number of tosses}} \text{(heads)}$$

_ _

(a) $p = 1.0$
(b) $p = 0.5$
(c) $p = 0.5$
(d) $p = 0.0$

36.   Draw a frequency distribution for the statistic p based on this mathematical analysis.

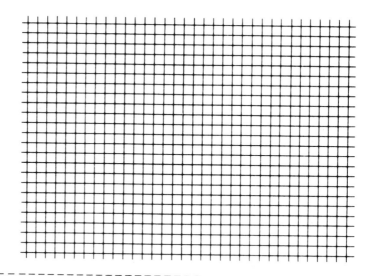

- -

Your distribution should look like this:

37. If you continue to take samples of two and compute the statistic p, you will find that the frequency distribution of the values of p comes closer and closer to matching the theoretical distribution you have just deduced. If you feel so inclined, take another 15 samples of two tosses and compute values of p. Compare the actual distribution of the 20 values of p with the theoretical distribution shown by the dark line on the graph below.

38. The distribution of a statistic is called a sampling distribution. The distributions of p you have been working with are _____ distributions.

- -

sampling

39. Which of the distributions below (a, b, or c) is a sampling distribution?
Which is a sample distribution?
Which is a population distribution?

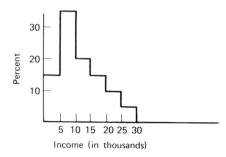

Figure a Distribution of incomes in a random sample of 1000 households.

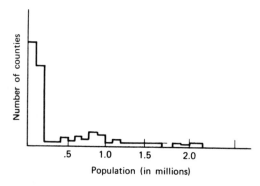

Figure b Distribution of county populations in 1950 Census data.

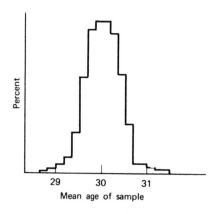

Figure c Distribution of mean age in 1000 samples of 20 individuals each,
drawn from a population of mean age 30.

Sampling distribution	(c)	This is the distribution of a statistic (the mean ages of 1000 samples).
Sample distribution	(a)	This is the distribution of observations in a sample (1000 random households).
Population distribution	(b)	This is the distribution of all possible observations in the population (all counties in the US).

40. The distribution of the individual observations in a sample is called a _____ distribution.

- -

sample

41. The distribution of all conceivable observations is called a _____ distribution.

- -

population

42. The distribution of a statistic is called a _____ distribution.

- -

sampling

43. The median weight of the 20,000 members of a national health club is 175 lb. For the sake of this example we assume that no one weighs precisely 175 lb. We should like to know what the probability is that three members, selected at random, all weigh more than 175 lb. To find out we can use a method similar to the one we used to analyze the coin-tossing problem. First, in the population we are dealing with (members of the club) what proportion weighs more than 175 lb?

$p =$ ____

- -

0.5. Remember, the median is the point in the distribution at which half the measurements are higher and half lower. So, if half the population weighs more than 175 lbs,

$$p = \frac{10,000}{20,000} = 0.5$$

44. We can use this information to construct a theoretical sampling distribution for samples of three drawn from this population. What are the possible outcomes of selecting the first member at random?

— —

Over 175
Under 175

45. Is either more likely than the other?

— —

No

46. If the population is large, will the first choice have any effect on the next choice?

— —

No

47. Draw a table like the one in frame 33 to show all possible results of selecting three members at random.

First selection	Second selection	Third selection

- -

First selection	Second selection	Third selection
Over 175	Over 175	Over 175
		Under 175
	Under 175	Over 175
		Under 175
Under 175	Over 175	Over 175
		Under 175
	Under 175	Over 175
		Under 175

48. Using the table, list all possible results of the three choices and compute the proportion that weighs over 175 lb for each possible result. That is, count the number of "successes" and divide by the number of trials.

	Result			p
(a)	Over	over	over	
(b)	Over	over	under	
(c)	Over	under	over	
(d)	Over	under	under	
(e)	Under	over	over	
(f)	Under	over	under	
(g)	Under	under	over	
(h)	Under	under	under	

- -

	Result			p	
(a)	Over	over	over	1.0	3/3
(b)	Over	over	under	0.67	2/3
(c)	Over	under	over	0.67	2/3
(d)	Over	under	under	0.33	1/3
(e)	Under	over	over	0.67	2/3
(f)	Under	over	under	0.33	1/3
(g)	Under	under	over	0.33	1/3
(h)	Under	under	under	0.00	0/3

49. Draw a sampling distribution for *p* under these conditions.

— —

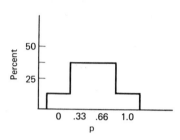

50. We are interested in how often we could expect to draw a sample of three members and find that all three weigh over 175 lb. What value of the statistic *p* would this situation correspond to?

— —

$p = 1.0$

51. Look at your sampling distribution for p. What percent of the time would you expect to draw a sample that gave you a p of 1.0?

_ _

12.5% of the time

THE BINOMIAL PROBABILITY DISTRIBUTION

The binomial probability distribution describes the sampling distribution of the statistic p for many possible values of P and many possible sample sizes.

52. Using the same sort of reasoning you have just applied to the coin tossing and weight problems, it is possible to work out sampling distributions for p not only for various sizes of sample but also for cases in which P is some other value than 0.5; for example, suppose you know that in a large population one out of four is a college graduate. What is P in this case?

_ _

$P = 0.25$ or $1/4$

53.

BINOMIAL PROBABILITIES

								P							
n	x	0.050	0.100	0.200	0.25	0.300	0.400	0.500	0.600	0.700	0.750	0.800	0.900	0.950	
2	0	0.902	0.810	0.640	0.563	0.490	0.360	0.250	0.160	0.090	0.063	0.040	0.010	0.002	
	1	0.095	0.180	0.320	0.375	0.420	0.480	0.500	0.480	0.420	0.375	0.320	0.180	0.095	
	2	0.002	0.010	0.040	0.063	0.090	0.160	0.250	0.360	0.490	0.563	0.640	0.810	0.902	
3	0	0.857	0.729	0.512	0.422	0.343	0.216	0.125	0.064	0.027	0.016	0.008	0.001	0.000	
	1	0.135	0.243	0.384	0.422	0.441	0.432	0.375	0.288	0.189	0.141	0.096	0.027	0.007	
	2	0.007	0.027	0.096	0.141	0.189	0.288	0.375	0.432	0.441	0.422	0.384	0.243	0.135	
	3	0.000	0.001	0.008	0.016	0.027	0.064	0.125	0.216	0.343	0.422	0.512	0.729	0.857	
4	0	0.815	0.656	0.410	0.316	0.240	0.130	0.062	0.026	0.008	0.004	0.002	0.000	0.000	
	1	0.171	0.292	0.410	0.422	0.412	0.346	0.250	0.154	0.076	0.047	0.026	0.004	0.000	
	2	0.014	0.049	0.154	0.211	0.265	0.346	0.375	0.346	0.265	0.211	0.154	0.049	0.014	
	3	0.000	0.004	0.026	0.047	0.076	0.154	0.250	0.346	0.412	0.422	0.410	0.292	0.171	
	4	0.000	0.000	0.002	0.004	0.008	0.026	0.062	0.130	0.240	0.316	0.410	0.656	0.815	
5	0	0.774	0.590	0.328	0.237	0.168	0.078	0.031	0.010	0.002	0.001	0.000	0.000	0.000	
	1	0.204	0.328	0.410	0.396	0.360	0.259	0.156	0.077	0.028	0.015	0.006	0.000	0.000	
	2	0.021	0.073	0.205	0.264	0.309	0.346	0.312	0.230	0.132	0.088	0.051	0.008	0.001	
	3	0.001	0.008	0.051	0.088	0.132	0.230	0.312	0.346	0.309	0.274	0.205	0.073	0.021	
	4	0.000		0.006	0.015	0.028	0.077	0.156	0.259	0.360	0.396	0.410	0.328	0.204	
	5								0.078		0.237	0.328	0.590	0.774	

These sampling distributions are listed in a table called a binomial probability table. A small excerpt from such a table is given above. To use the table you must know what P is and what the sample size is. (The usual symbol for sample size is n). The sampling distribution is given in terms of the number of "successes" rather than values of the statistic p; for example, in the case of tossing coins P was 0.5 and the sample size n was 2. The appropriate sampling distribution is circled in the table. The number of "successes," that is, the number of times the coin comes up heads, is indicated by x. According to the circled portion of the table, how frequently will there be two successes in a sample of two? _____

- -

0.25; that is, 25% of the time

54. For the weight problem, what was P? Look back at frame 43 if you do not recall the problem.

- -

$P = 0.5$

55. What was n?

_____

$n = 3$ (n is the sample size)

56. Circle the appropriate sampling distribution in the table in frame 53.

_____

BINOMIAL PROBABILITIES

n	x	0.050	0.100	0.200	0.250	0.300	0.400	0.500	0.600	0.700	0.750	0.800	0.900	0.950
2	0	0.902	0.810	0.640	0.563	0.490	0.360	0.250	0.160	0.090	0.063	0.040	0.010	0.002
	1	0.095	0.180	0.320	0.375	0.420	0.480	0.500	0.480	0.420	0.375	0.320	0.180	0.095
	2	0.002	0.010	0.040	0.063	0.090	0.160	0.250	0.360	0.490	0.563	0.640	0.810	0.902
3	0	0.857	0.729	0.512	0.422	0.343	0.216	0.125	0.064	0.027	0.016	0.008	0.001	0.000
	1	0.135	0.243	0.384	0.422	0.441	0.432	0.375	0.288	0.189	0.141	0.096	0.027	0.007
	2	0.007	0.027	0.096	0.141	0.189	0.288	0.375	0.432	0.441	0.422	0.384	0.243	0.135
	3	0.000	0.001	0.008	0.016	0.027	0.064	0.125	0.216	0.343	0.422	0.512	0.729	0.857
4	0	0.815	0.656	0.410	0.316	0.240	0.130	0.062	0.026	0.008	0.004	0.002	0.000	0.000
	1	0.171	0.292	0.410	0.422	0.412	0.346	0.250	0.154	0.076	0.047	0.026	0.004	0.000
	2	0.014	0.049	0.154	0.211	0.265	0.346	0.375	0.346	0.265	0.211	0.154	0.049	0.014
	3	0.000	0.004	0.026	0.047	0.076	0.154	0.250	0.346	0.412	0.422	0.410	0.292	0.171
	4	0.000	0.000	0.002	0.004	0.008	0.026	0.062	0.130	0.240	0.316	0.410	0.656	0.815
5	0	0.000	0.390	0.390	0.237	0.168	0.078	0.031	0.010	0.007	0.001			
			0.008	0.000	0.000									

57. You are told that in the population of all doctors 9 out of 10 recommend Potter's Pills. Assuming this is true, you want to know the probability of choosing two doctors at random and finding that neither of them recommends Potter's Pills. What are P and n for this case?

_____

$P = 0.9 \left(\frac{9}{10}\right)$
$n = 2.0$

58. What value of x corresponds to the case in which neither of the two doctors recommends Potter's Pills? (A doctor who recommended Potter's Pills would be a "success.")

– –

$x = 0$

59. How likely is it that neither doctor will recommend Potter's Pills? Use the table in frame 53.

– –

The probability of this event is 0.010; that is, it should happen 1% of the time.

60. Let us make the problem more complicated. Suppose you plan to choose four doctors at random and you want to know the probability that no more than two will recommend Potter's Pills. To come up with an answer you must take into account the case in which none of them recommends Potter's Pills (zero "successes"), the case in which only one recommends Potter's Pills and three do not (one "success"), and the case in which two recommend Potter's Pills and two do not (two "successes"). You will find the probability of each of these cases in the sampling distribution; then you will add the probabilities for each of these cases. What are P and n for this problem? What is the probability that one of these three situations will exist?

– –

$P = 0.9$, $n = 4$. The probability that x is less than or equal to 2 is 0.053. To find this answer you had to add the probabilities of the three situations of interest. The probability that $x = 0$ is too small to enter in the table. The probability that $x = 1$ is 0.004 and the probability that $x = 2$ is 0.049.

61. The total of all the probabilities in a sampling distribution is always 1.0. If the probability of no, one, or two recommendations for Potter's Pills is 0.053, what is the probability of three or four recommendations?

- -

0.947 (1.000–0.053).

62. Five percent of the population of the United States possesses a particular genetic trait. You want to know the chances of finding at least one person who has this trait in a random sample of 15. Use Table III in the back of the book.

- -

$p = 0.05$, $n = 15$. The probability is 0.537, or about 54%. The easiest way to find the answer is to find the probability of $x = 0$, that is, the probability that no one in the sample will have the trait. The probability that no one will have the trait is 0.463; therefore the probability that at least one person will have the trait is $1.000 - 0.463 = 0.537$.

You could also have added up the probabilities for $x = 1$, $x = 2$, $x = 3$, and so on. The sum of these probabilities from the table is 0.538. Because the table entries are rounded off, you will find occasional minor discrepancies like this.

63. When you use the binomial probability table, you are making an important assumption that you should be aware of. You are assuming that the observations in your sample are random and independent; that is, you are assuming that every observation in the population has an equal chance of being chosen at any step in the sampling process. Suppose you are going to choose three people out of a population of six that consists of three men and three women. You choose to select them one at a time. On the first draw you choose a man. Does this affect the chances of your choosing a man on the second draw?

Yes, because there are now substantially fewer men in the remaining population. For the first draw $P = \frac{3}{6} = 0.5$; for the second draw $P = \frac{2}{5} = 0.4$.

64. You have a jar containing six marbles, three red and three blue. You draw one marble at random and find that it is blue. You record this observation and replace the marble in the jar. Has your first observation affected the chances of choosing a blue marble on the second draw?

No, because you replaced the blue marble.

65. Can you use the binomial probability table to analyze the marble problem?

Yes. The observations are independent. One observation does not affect the probability of the others.

66. Can you use the binomial probability table to analyze the problem in frame 63?

No. The observations are not independent.

Procedures for dealing with this problem exist, but they are beyond the scope of this book.

67. As a rule of thumb, if the population is at least 20 times as large as the sample, you can disregard this effect of not being able to include one individual in your sample twice. If you are choosing a sample of six out of a population of 500, can you use the binomial probability table?

_ _

Yes

68. If you are choosing a sample of six out of a population of 75, can you use the binomial probability table?

_ _

No

69. You recall that if nine out of 10 doctors recommend Potter's Pills the probability of selecting two doctors at random who do not recommend them is 0.010. You go to a nearby medical school and ask the first two doctors you meet whether they recommend Potter's Pills. Have all members of the population an equal chance of being chosen?

_ _

No. The chance of choosing doctors connected with that medical school is much greater than the chance of choosing other doctors.

70. Can you say that the probability of two "no" answers is 0.010?

_ _

No

71. What assumption do you make when you use the binomial probability table?

_ _

All members of the population have an equal chance of being selected at any point in the sampling procedure.

THE NORMAL DISTRIBUTION

It is possible to deduce the sampling distributions of a number of other statistics mathematically. This book will not attempt to explain the mathematical reasoning involved in deducing these sampling distributions. Instead, our focus will be on how and why sampling distributions are useful in drawing conclusions about populations and samples.

72. One of the most useful sampling distributions is the one for the mean of a large sample of measurements. This distribution is called the normal distribution. Before explaining it, let's review the symbols we will be using.

μ is the symbol for the _____ of the (sample/ population).

σ is the symbol for the _____ of the (sample/ population).

– –

mean of the population
standard deviation of the population

73. What does n stand for?

– –

Size of the sample

74. μ is a (parameter/statistic)?

– –

parameter

75. The sample mean is a (parameter/statistic)?

– –

statistic

76. We represent the sample mean by the symbol $\bar{x}$; for example, if the mean height of all women in the United States is 5 ft 4 in., and a sample of 10 women has a mean height of 5 ft 9 in.

$\mu =$ _____

$\bar{x} =$ _____

- -

$\mu = 5$ ft 4 in.
$\bar{x} = 5$ ft 9 in.

77. If we take a large number of samples and compute an $\bar{x}$ for each sample, we can also compute a mean for the $\bar{x}$'s as well as a standard deviation. We call these values $\mu_{\bar{x}}$ and $\sigma_{\bar{x}}$. Complete the following table:

	Sample	Sampling Distribution of Mean	Population
Mean			μ
Standard deviation	s^*		

- -

	Sample	Sampling Distribution of Mean	Population
Mean	$\bar{x}$	$\mu_{\bar{x}}$	μ
Standard deviation	s	$\sigma_{\bar{x}}$	σ

*You will learn more about s in Chapter 3.

78.

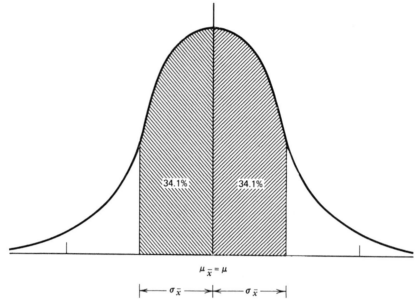

$$\mu_{\bar{x}} = \mu$$

Normal distribution.

According to what is called the central limit theorem, the sampling distribution of $\bar{x}$ tends to have the shape of the normal distribution curve shown above, with

$$\mu_{\bar{x}} = \mu \quad \text{and} \quad \sigma_{\bar{x}} = \frac{\sigma}{\sqrt{n}}$$

When the sample size n is 30 or more, the sampling distribution is almost exactly like the illustration. Suppose we took a great number of samples of 36 individuals from a population with a mean age of 40 and a standard deviation of six years. What would the mean of the sampling distribution be?

$$\mu_{\bar{x}} = \underline{\quad} = \underline{\quad}$$

- -

$$\mu_{\bar{x}} = \mu = 40$$

79. What would the standard deviation of the sampling distribution be?

$\sigma_{\overline{x}} = $ _____

_ _

$\sigma_{\overline{x}} = \dfrac{\sigma}{\sqrt{n}} = \dfrac{6}{\sqrt{36}} = 1$

80. According to the illustration, what percent of the time would you expect the mean of one sample to be between 40 and 41?

_ _

34.1%

81. What percent of the time would you expect the mean of one sample to be between 39 and 41?

_ _

68.2%

82.

According to the central limit theorem, the shape of the sampling distribution of $\bar{x}$ will be a normal curve, no matter what the shape of the population distribution. The illustration above shows the population distribution for time to solve a particular puzzle for 6000 individuals. A large number of samples of 49 individuals each are selected and $\bar{x}$ is computed for each sample. Which of the illustrations below will most resemble the sampling distribution of $\bar{x}$?

(a)

(b)

(c)

– –

(c). No matter what the population distribution or the sample distributions look like, the sampling distribution of the statistic $\bar{x}$ will always have this bell shape.

83. Remember that $\sigma_{\bar{x}} = \dfrac{\sigma}{\sqrt{n}}$.

What will the mean of the sampling distribution be? What will its standard deviation be?

- -

$\mu_{\bar{x}} = \mu = 75$

$\sigma_{\bar{x}} = \dfrac{\sigma}{\sqrt{n}} = 64/7 = 9.14$

84. The shape of the sampling distribution of $\bar{x}$ is called a normal distribution. In this normal distribution there is always 34.1% of the observations between $\mu_{\bar{x}}$ and $\mu_{\bar{x}} + \sigma_{\bar{x}}$, 34.1% of the observations between $\mu_{\bar{x}} - \sigma_{\bar{x}}$ and $\mu_{\bar{x}}$, 47.7% of the observations between $\mu_{\bar{x}}$ and $\mu_{\bar{x}} + 2\sigma_{\bar{x}}$, and so on. The exact shape

of the distribution is completely defined by the two parameters _____ and _____.

- -

$\mu_{\bar{x}}$ and $\sigma_{\bar{x}}$

85. If we know how many standard deviations away from the mean an observation is, we can tell how likely it is to occur in random sampling. For this reason it is convenient to convert our measurements into z scores. A z score is simply the number of standard deviations away from the mean a particular measurement is located; for example, a score one standard deviation above the mean has a z score of $+1$; a score one and one-half standard deviations below the mean has a z score of -1.5.

If a distribution has a mean of 15 and a standard deviation of 2, what z score corresponds to a raw score of 19? What z score corresponds to a raw score of 14?

- -

$z = +2$; that is, two standard deviations above the mean

$z = -0.5$; that is, one-half standard deviation below the mean

86. The formula for converting any measurement to a z score is

$$z = \frac{\bar{x} - \mu}{\sigma_{\bar{x}}}$$

If $\mu = 15$ and $\sigma_{\bar{x}} = 2$, what z score corresponds to a raw score of 20?

$$z = \frac{20 - 15}{2} = 2.5$$

87. The table that gives the normal frequency distribution is usually called "areas under the normal curve" and is organized in terms of z scores. If you want to know the probability of a particular $\bar{x}$ from a known population, you must first compute a _____ .

----

z score

88.   AREAS UNDER THE NORMAL CURVE

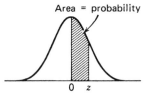

Area = probability

An entry in the table is the proportion under the entire curve which is between $z = 0$ and a positive value of z. Areas for negative values of z are obtained by symmetry.

Second decimal place of z

z	.00	.01	.02	.03	.04	.05	.06	.07	.08	.09
0.0	.0000	.0040	.0080	.0120	.0160	.0199	.0239	.0279	.0319	.0359
0.1	.0398	.0438	.0478	.0517	.0557	.0596	.0636	.0675	.0714	.0753
0.2	.0793	.0832	.0871	.0910	.0948	.0987	.1026	.1064	.1103	.1141
0.3	.1179	.1217	.1255	.1293	.1331	.1368	.1406	.1443	.1480	.1517
0.4	.1554	.1591	.1628	.1664	.1700	.1736	.1772	.1808	.1844	.1879
0.5	.1915	.1950	.1985	.2019	.2054	.2088	.2123	.2157	.2190	.2224
0.6	.2257	.2291	.2324	.2357	.2389	.2422	.2454	.2486	.2517	.2549
0.7	.2580	.2611	.2642	.2673	.2703	.2734	.2764	.2794	.2823	.2852
0.8	.2881	.2910	.2939	.2967	.2995	.3023	.3051	.3078	.3106	.3133
0.9	.3159	.3186	.3212	.3238	.3264	.3289	.3315	.3340	.3365	.3389
1.0	.3413	.3438	.3461	.3485	.3508	.3531	.3554	.3577	.3599	.3621
1.1	.3643	.3665	.3686	.3708	.3729	.3749	.3770	.3790	.3810	.3830
1.2	.3849	.3869	.3888	.3907	.3925	.3944	.3962	.3980	.3997	.4015
1.3	.4032	.4049	.4066	.4082	.4099	.4115	.4131	.4147	.4162	.4177
1.4	.4192	.4207	.4222	.4236	.4251	.4265	.4279	.4292	.4306	.4319
1.5	.4332	.4345	.4357	.4370	.4382	.4394	.4406	.4418	.4429	.4441

The table of areas under the normal curve gives you the probability of a sample mean between μ and $\mu + z$; for example, if you want to know the probability of a sample mean between μ and $\mu + 1.0\sigma$, you will look in the table under $z = 1.00$. The probability in this case is 0.3413. What is the probability of a sample mean between μ and $\mu + 1.5\sigma$?

- -

0.4332

89. A population has a mean of 20 and a standard deviation of 4. You plan to choose a sample of 64 at random. What is the probability of a sample mean between 20 and 21? To answer this question you must find the mean and standard deviation of the sampling distribution and use them to convert 21 to a z score.

What is the mean of the sampling distribution? What is the standard deviation of the sampling distribution? What z score corresponds to 21?

- -

$$\mu_{\bar{x}} = 20$$

$$\sigma_{\bar{x}} = \frac{\sigma}{\sqrt{n}} = 0.5$$

$$z = \frac{21 - \mu}{0.5} = \frac{1}{0.5} = 2.0$$

90. Use the table on page 269 to find the probability of a sample mean between 20 and 21.

- -

0.477

91. Often you will have to add or subtract values you find in the table to answer a question; for example, what is the probability of a sample mean between 19 and 21? Break this down into the probability of an $\bar{x}$ between 20 and 21 and the probability of an $\bar{x}$ between 20 and 19. The answer is _____ .

- -

0.9544

92. It helps to remember that the total of all probabilities in a sampling distribution is always 1.0000. If the probability of a sample mean between 19 and 21 is 0.9544 (i.e, 95% of the time), what is the probability of a sample mean that is not between 19 and 21 (either less than 19 or more than 21)?

- -

0.045 6   (1.000–0.9544), that is, 5%

93. The total of all probabilities on one side of the mean is 0.5000. A sample mean will be above the mean of the sampling distribution half the time. What is the probability of a sample mean greater than 21?

_ _

0.0228 (0.5000–0.4772)

94. Reference formulas

$$\mu_{\bar{x}} = \mu$$

$$\sigma_{\bar{x}} = \frac{\sigma}{\sqrt{n}}$$

$$z = \frac{\bar{x} - \mu}{\sigma_{\bar{x}}}$$

An airline has determined that the mean weight of a passenger's baggage is 28.5 kg, with a standard deviation of 5 kg. If we take random samples of 100 passengers, how often can we expect to find a sample mean over 30 kg?

_ _

Population distribution

$\mu = 28.5$

$\sigma = 5.0$

Sampling distribution

$\mu_{\bar{x}} = 28.5$

$\sigma_{\bar{x}} = \dfrac{5}{\sqrt{100}} = 0.5$

z score corresponding to 30 kg

$z = \dfrac{30 - 28.5}{0.5} = 3.0$

Probability of $\bar{x}$ between 28.5 and 30

(from table) 0.4987

Probability of $\bar{x}$ over 30

0.0013 (0.5000–0.4987), that is, about once in a thousand times

95. The airline's planes hold 100 passengers and, to the delight of the stockholders, are always full. Can we consider a plane load of passengers as a random sample of 100 airline passengers? Has every member of the population an equal chance of being selected at each stage of the sampling process?

- -

No, a planeload is not a random sample. For example, the presence of one member of a group tour or a family on a plane greatly increases the probability of other members of the same group being on the plane.

96. According to the central limit theorem, the shape of the sampling distribution of $\bar{x}$ for large samples tends to be normal, with $\mu_{\bar{x}} = $ ____ and $\sigma_{\bar{x}} = $ _____ .

- -

$\mu_{\bar{x}} = \mu$

$\sigma_{\bar{x}} = \dfrac{\sigma}{\sqrt{n}}$

97. The shape of the sampling distribution of $\bar{x}$ for large samples (n greater than 30) is called _____ .

----

normal

98. For large samples (n greater than 30) the sampling distribution of $\bar{x}$ tends to be normal, with mean and standard deviation related to the population parameters as follows:

$$\mu_{\bar{x}} = \mu$$

$$\sigma_{\bar{x}} = \frac{\sigma}{\sqrt{n}}$$

This is a statement of the _____ theorem.

----

central limit

99. On a separate sheet of paper, state the central limit theorem in your own words.

Your answer should have included these points:
(a) The theorem has to do with the sampling distribution of the means of large samples (greater than 30).
(b) The shape of this sampling distribution tends to be normal.
(c) The mean of this distribution is equal to the population mean.
(d) The standard deviation of this distribution equals the population standard deviation divided by the square root of the sample size.

REVIEW PROBLEMS

If you have successfully completed this chapter, you can now use information about a population to make predictions about samples. You can:

● tell the difference between a population and a sample;
● use a binominal probability table to predict sample proportions, given the population proportion;
● use a normal distribution table to predict sample means, given the mean and standard deviation of the population.

Now try these review problems. Table I on pp. 253–256 lists any formulas you may need for reference.

1. Write the term that corresponds to each of the following descriptions:

   (a) A group of observations chosen at random to represent a larger group of possible observations.

   (b) All possible observations of a given type.

   (c) A number that summarizes a population of observations.

   (d) A number that summarizes a sample of observations.

2. You have two complete decks of cards with no jokers. You shuffle the cards thoroughly and draw five cards. What is the probability that you will draw four or more clubs?

3. You take only the face cards and aces from a deck of cards and discard the others. Then you shuffle the cards thoroughly and draw two cards. Can you use the binomial probability table to calculate the probability of drawing two clubs? Why?

4. The mean diameter of marbles manufactured at a particular toy factory is 0.850 cm with a standard deviation of 0.010 cm. What is the probability of selecting a random sample of 100 marbles that has a mean diameter greater than 0.851 cm?

Answers

To review a problem, study the frames indicated after the answer.

1. (a) sample
    (b) population
    (c) parameter
    (d) statistic
    See frames 1 to 17.
2. The population is 104 and the sample size 5, so you may use the binomial probability table even though the cards are not replaced.

    $P = 0.25$ (there are equal numbers of the four suits).
    $n = 5$ (you will draw five cards).
    $x = 4, 5$ (you are interested in four or more successes).

    From the binomial probability table you find these values: 0.015, 0.001. The probability of drawing four or more clubs is 0.016; it would happen about 1.6% of the time. *See frames 52 to 71.*
3. No. Your deck (the population) consists of only 16 cards. Since this is less than 20 times the sample size, you cannot use the binomial probability table unless you replace the card after each individual draw. *See frames 63 to 68.*
4. $\mu_{\bar{x}} = 0.85$
    $\sigma = 0.01$

    $$\sigma_{\bar{x}} = \frac{\sigma}{\sqrt{n}} = \frac{0.01}{10} = 0.001$$

    $$z = \frac{\bar{x} - \mu_{\bar{x}}}{\sigma_{\bar{x}}} = \frac{0.851 - 0.850}{0.001} = \frac{0.001}{0.001} = 1.0$$

    From the normal probability table the probability of $\bar{x}$ between 0.850 and 0.851 is 0.341. The probability of $\bar{x}$ greater than 0.851 is $0.500 - 0.341 = 0.159$, or 15.9%. *See frames 72 to 99.*

CHAPTER THREE
Estimating

Often we need to estimate the characteristics of a population on the basis of information about a sample. For example, you may want to estimate the mean height of all the trees in the forest without having to measure every tree. Common sense suggests that you can get a pretty good idea of the height of all trees in the forest by measuring a random sample of trees. In fact, by using sampling-distribution tables, you can translate the description "pretty good" into a mathematical statement of your level of confidence. This chapter will teach you how to estimate population parameters on the basis of data from a sample and how to establish mathematically defined confidence intervals for these estimates.

When you have completed this chapter, you will be able to:

- estimate μ, P, and σ;
- establish confidence intervals for P;
- establish confidence intervals for μ.

ESTIMATION

Often we use a sample as a basis for estimating the value of population parameters. Even though we know that the sample statistic is not identical to the population parameter, the statistic is our best estimate of the parameter.

1. If the mean weight of 30 cartons of "large" eggs is 30 oz, what would your best estimate be of the mean weight of all cartons of "large" eggs?

- -

30 oz

2. In this case you used the statistic ___ as an estimate of the parameter ___.

- -

$\bar{x}$

μ

3. A marketing research organization reports that 18% of the television audience was watching "The Return of the Creature from the Black Lagoon" on a given evening. Do you think this proportion is a parameter or a statistic?

– –

A statistic. This report is based on a sample of television watchers.

4. In this case p, a (parameter/statistic), is being used as an estimate of P, a (parameter/statistic).

– –

statistic
parameter

5. When we wish to estimate the standard deviation of a population on the basis of a sample distribution, we encounter a problem. The number that you obtain by applying the formula for σ to the sample data is not a good estimate of the population standard deviation. Instead, a slightly different formula must be used.

An explanation would be too long to present here, but with small samples the formula for σ tends to underestimate the standard deviation of the population. That is, on the average, the standard deviation of a sample tends to be smaller than the standard deviation of the population.

The two formulas are:

Standard deviation of population	Estimate of population standard deviation based on sample data
$\sigma = \sqrt{\dfrac{\Sigma(x-\mu)^2}{n}}$	$s = \sqrt{\dfrac{\Sigma(x-\overline{x})^2}{n-1}}$

The symbol for population standard deviation is ——.

– –

σ

6. The symbol for the estimate of population standard deviation is——.

– –

s

7. In these formulas, σ is a (parameter/statistic); s is a (parameter/statistic). The formula for σ contains the expression n where the formula for s contains the expression _____ and μ where the formula for s contains ___ .

- -

σ is a parameter; s is a statistic

$n-1$

$\bar{x}$

8. Which formula contains the expression $n - 1$?

- -

$$s = \sqrt{\frac{\Sigma(x - \bar{x})^2}{n - 1}}$$

9. What is the formula for σ?

- -

$$\sigma = \sqrt{\frac{\Sigma(x - \mu)^2}{n}}$$

10. What is the formula for s?

- -

$$s = \sqrt{\frac{\Sigma(x - \bar{x})^2}{n - 1}}$$

11. If you wish to estimate the standard deviation of the weight of all cartons of "large" eggs on the basis of a sample of 30 cartons, which formula would you use (write out the whole formula)?

- -

$$s = \sqrt{\frac{\Sigma(x - \overline{x})^2}{n - 1}}$$

12. If 20 cartons of eggs constitute a population and you wish to compute the standard deviation of this population, which formula will you use (write out the whole formula)?

_ _

$$\sigma = \sqrt{\frac{\Sigma(x - \mu)^2}{n}}$$

13. The following represent the number of words in the speaking vocabulary of five children. Describe this group in terms of mean and standard deviation. 100, 100, 300, 400, 600.

_ _

$\mu = 300$

$$\sigma = \sqrt{\frac{180000}{5}} = 189.7$$

Since you are not trying to estimate the parameters of some larger population, you should have used the formula for σ.

14. Five children of a given age in an appropriately selected sample have the following number of words in their speaking vocabularies. What is the best estimate you can make of the mean and standard deviation for all children of that age.

100, 100, 300, 400, 600.

- -

$\bar{x} = 300$

$$s = \sqrt{\frac{180000}{4}} = 212.1$$

Since this is an estimation problem, you should have used the formula for s. (As you will see, with such a small sample these estimates are not very reliable; but they are the best estimates you can make.)

15. If you compute both s and σ for the same data, which will be larger?

- -

s. The divisor is smaller.

16. With a sample size of 1000, would you expect the difference between σ and s to be substantial?

- -

No. The difference between a given number divided by 1000 and the same number divided by 999 is not very great.

17. With a sample size of 4, would you expect the difference between σ and s to be substantial?

_____

Yes. There is a substantial difference between $x/3$ and $x/4$.

18. What statistic is used as an estimate of P?

_____

p

19. What statistic is used as an estimate of σ?

_____

s

20. What statistic is used as an estimate of μ?

_____

$\overline{x}$

CONFIDENCE INTERVALS FOR μ

The amount of confidence you can put in an estimate of a parameter varies. In some cases you can say that it is virtually inconceivable for your estimate to be wrong by an amount that has any practical significance. In other cases you must consider your estimate, at best, a "ball-park" figure, an indicator only of the approximate magnitude of the parameter of interest. One useful way to deal with this problem is to establish a confidence interval for your estimate; for example, you might say "the mean number of characters per message in our data communications system is between 307 and 313 with 98% confidence."

21. To clarify how you can establish a confidence interval, and what it means, let us consider a somewhat artificial example. Suppose you are trying to estimate the mean of a population in a situation in which you know the population variability.

    A factory has two machines that cut spaghetti and noodles into appropriate lengths for packaging. Both machines can be adjusted to cut any given length. The variability of the two machines differs, however. The standard deviation of pasta cut on machine A is 0.1 cm. The standard deviation of pasta cut on machine B is 1.0 cm.

    If the factory manager wants all noodles in a batch to be as nearly as possible the same length, which machine will he use to cut them?

    _

    Machine A, because its output is less variable.

22. Now consider what would happen if you took a number of samples of pasta cut by machine A and computed the mean of each sample. Suppose for example that the machine is set to cut pasta 30 cm in length and you take a number of 36-noodle samples. We can predict the sampling distribution for machine A on the basis of the central limit theorem. Complete the following table:

    Machine A

    Population $\mu = 30$ cm
                  $\sigma = 0.1$ cm

    Samples of 36 $\mu_{\bar{x}} =$ $\mu =$ _____

                $\sigma_{\bar{x}} = \dfrac{\sigma}{\sqrt{n}} =$ _____

    _

Machine A

$\mu_{\bar{x}} = 30$ cm

$\sigma_{\bar{x}} = \dfrac{0.1}{\sqrt{36}} = \dfrac{0.1}{6} = 0.0167$ cm

23. According to the normal distribution table, how often will $\bar{x}$ be within 0.0167 cm of μ?

- -

0.6826. Roughly 68% of the time. Since $\sigma_{\bar{x}} = 0.0167$, $z = 1.0$. The entry in the normal distribution table for $z = 1.0$ is 0.3413. You are interested in values either above or below μ, so you must double the value in the table.

24. "Under the circumstances described ($\sigma = 0.1$, $n = 36$) $\bar{x}$ will be within 0.0167 cm of μ 68% of the time." True or false?

- -

True

25. Now, suppose that μ is not known. If σ and n remain unchanged, this statement will be true, no matter what the value of μ. If μ is unknown, there is a 68% chance that we will be correct in predicting that it is within 0.0167 cm of whatever $\bar{x}$ we compute from a sample; for example, if we obtain an $\bar{x}$ of 20 from our sample, there is a 68% chance that μ is between $20 + 0.0167$ and $20 - 0.0167$.

We can say that μ is between 19.9833 and _____ with 68% confidence.

- -

20.0167

26. If you are going to draw practical conclusions from your estimates, you will probably want a higher confidence level than 68%; for example, you might want to take only a 5% risk of making an incorrect estimate. Then you would choose a confidence level of _____.

- -

95%

27. Assume that you have taken a sample of 36 noodles cut on machine A and have found that the mean length of the 36 noodles is 25 cm. You now want to determine the exact setting of the machine with 95% confidence. Here is how to establish a 95% confidence interval:

(a) Decide what confidence level you will use. We chose 95%.

(b) Divide the confidence level by two to find what percentage of the sampling distribution must be included on each side of the mean. To include 95% you must include 47.5% on each side of the mean.

(c) Look in the normal distribution table to find the z score that includes the appropriate percentage. We find that 0.475 corresponds to a z score of 1.96.

(d) The z score tells you how many standard deviations on either side of the mean you must go to include the desired percentage. You must multiply it by the standard deviation of the sampling distribution to obtain a value in centimeters.

$$1.96 \times 0.0167 = .033 \text{ cm}$$

(e) Add the value in centimeters to the sample mean to obtain the upper limit of the confidence interval and subtract it from the sample mean to obtain the lower limit. Since the sample mean was 25.000, the upper limit is _____. The lower limit is _____.

– –

25.033
24.967

28. Now assume that you want to compute a 99% confidence interval on the basis of the same sample:

$x = 25.000$ cm
$\sigma = 0.1$ cm
$n = 36$
$\sigma_{\bar{x}} = \dfrac{\sigma}{\sqrt{n}} = 0.0167$

Compute a 99% confidence level for μ. Refer to frame 27, if you wish, as you complete these steps.

(a) What is the confidence level?

(b) What percent on each side of the mean?

(c) What z score?

(d) How many centimeters?

(e) What are the upper and lower limits?

– –

(a) 99%
(b) 49.5%
(c) 2.58
(d) 0.043 cm (2.58 × 0.0167)
(e) 24.957–25.043

29. z_0 is the symbol used to indicate the z score that corresponds to your choice of a confidence level; for example, with a 95% confidence level $z_0 = 1.96$. With a 99% confidence level $z_0 = $ _____.

_ _

2.58

30.
> The formula for a confidence interval for μ is $\bar{x} \pm z_0\left(\dfrac{\sigma}{\sqrt{n}}\right)$.

The symbol $\pm$ is read "plus or minus." In this formula it tells you to add to find the upper limit and subtract to find the lower limit. Using this formula as a reminder, compute the 99% confidence interval for the setting of machine B on the basis of the following information:

$\bar{x} = 25$
$\sigma = 1.0$
$n = 36$

μ is between _____ and _____ with 99% confidence.

_ _

24.57 and 25.43

31. What z_0 corresponds to a confidence level of 90%?

_ _

1.65

CONFIDENCE INTERVALS FOR μ WHEN σ IS UNKNOWN

When you do not know the population standard deviation, you can still calculate a confidence interval for μ on the basis of sample statistics. To do this, you must know both the mean and the standard deviation of the sample.

32. In the cases we have discussed so far you are estimating μ, but you know the exact value of the population parameter σ. Do you think this would be a common state of affairs?

- -

No. You are most likely to encounter situations in which both μ and σ are unknown.

33. If the sample size is 30 or more, you can obtain a close estimate of the true confidence interval by substituting an estimate for σ. What is the estimate for σ?

- -

s

34.   Reference formulas:

$$\text{Confidence interval for } \mu: \overline{x} \pm z_0 \frac{\sigma}{\sqrt{n}}$$

$$\text{Estimate of } \sigma: s = \sqrt{\frac{\Sigma(x - \overline{x})^2}{n - 1}}$$

You are studying the time it takes for an air-traffic controller to read a complex radar display and react to it under certain standardized conditions. By conducting one trial with each of 36 randomly selected controllers you obtain the following statistics.

$\overline{x} = 3.15$ seconds, $\Sigma(x - \overline{x})^2 = 20.00$, $n = 36$.

What is the 95% confidence interval for mean time to read and react to the display?

- -

(a)   95% confidence interval
(b)   47.5% on each side of the mean
(c) $z_0 = 1.96$
(d) $s = \sqrt{\dfrac{\Sigma(x - \overline{x})^2}{n - 1}} = \sqrt{\dfrac{20}{35}} = 0.755,$ $z_0 \dfrac{s}{\sqrt{n}} = 1.96\left(\dfrac{0.755}{6}\right) = 0.25$

(e)   The confidence interval is 3.15 ± 0.25;
       μ is between 2.90 and 3.40 with 95% confidence.

35. The normal probability table gives the sampling distribution of x for samples of 30 or more. Can you use it to establish a confidence interval for μ on the basis of a sample of 10?

— —

No

36. To use the normal probability table for establishing a confidence interval for μ you must have a sample size of at least _____.

— —

30

37. If your sample is smaller than 30, it is still possible to establish a confidence interval for μ, provided you can assume the population distribution is normal; for example, suppose in your study of air-traffic controllers you are able to observe only samples of 20. The graphs below summarize the data you obtain for three different tasks.

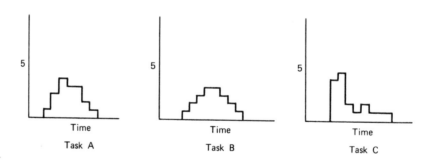

You are prepared from experience to assume that the population distributions are normal unless the data make this assumption unlikely. Would these data make you assume that the population distribution for any of these tasks is not normal? If so, which one(s)?

— —

Task C. The sample distribution is so skewed that it seems unlikely to have come from a normally distributed population. (Although the sample distribution for Task A is also slightly skewed, you are quite likely to draw a slightly skewed sample of 20 from a normally distributed population.)

38. For samples larger than 30 must you assume that the population is normally distributed in order to establish a confidence interval for μ?

_ _

No

39. For samples smaller than 30 must you assume that the population is normally distributed in order to establish a confidence interval for μ?

_ _

Yes

40. Can you establish a confidence interval for μ on the basis of a sample of 15 from a population that is not normally distributed?

_ _

No

41. When you establish a confidence interval for μ for a small, normally distributed sample, you must use a sampling distribution called "Student's t." This distribution is similar to the normal distribution, but its exact shape depends on the size of the sample. The procedure is the same except that

(a) instead of using a value of z_0 from the normal probability table, you must use a value of t_0 from a t table,

(b) you must always use s, even if σ is known;

for example, let us establish a 99% confidence interval for the mean time of task B in frame 37.

$$\overline{x} = 3.00 \qquad s = 0.53 \qquad n = 20$$

From the table we find that t_0 for 99% of a sample of 20 is 2.86. (Patience, you will learn how to read the table in a moment.)

The formula for the confidence interval of a small, normally distributed sample is

$$\overline{x} \pm t_0 \frac{s}{\sqrt{n}}$$

What is the confidence interval in this case?

- -

$\overline{x} \pm t_0 (s/\sqrt{n}) = 3.00 \pm 2.86 \, (0.53/\sqrt{20}) = 3.00 \pm 0.34$; μ is between 2.66 and 3.34 with 99% confidence.

CRITICAL POINTS OF THE t DISTRIBUTION

The first column lists the number of degrees of freedom (ν). The headings of the other columns give probabilities (P) for t to exceed the entry value. Use symmetry for negative t values.

P df	.10	.05	.025	.01	.005
1	3.078	6.314	12.706	31.821	63.657
2	1.886	2.920	4.303	6.965	9.925
3	1.638	2.353	3.182	4.541	5.841
4	1.533	2.132	2.776	3.747	4.604
5	1.476	2.015	2.571	3.365	4.032
6	1.440	1.943	2.447	3.143	3.707
7	1.415	1.895	2.365	2.998	3.499
8	1.397	1.860	2.306	2.896	3.355
9	1.383	1.833	2.262	2.821	3.250
10	1.372	1.812	2.228	2.764	3.169

42. The above is an excerpt from a t table. Each line represents the distribution of t for a particular sample size. The df (degree of freedom) column on the left allows you to select the correct distribution for your sample size. The degrees of freedom in a single sample are equal to $n-1$. Thus the degrees of freedom for a sample of eight are _____ .

_____

Seven

43. The degrees of freedom for a sample of 10 are _____ .

_____

Nine

44. The headings at the top of the table give the probabilities that the mean is outside the upper or lower limit set by the indicated values of t; for example, if you are setting a 95% confidence interval, you will want the probability of a mean outside the upper limit to be only 2.5%. (Of course, the probability of a mean outside the lower limit will also be 2.5%.) You will therefore choose a value of t from the column headed 0.025. What column will you use if you are setting a 90% confidence interval?

‒ ‒

The column headed 0.05

45. You want to look up t_0 for a 98% confidence interval for a sample of six. What value of df will you use?

‒ ‒

five (df $= n - 1$)

46. What column will you use?

‒ ‒

0.01

47. Circle the correct entry in the table at frame 42.

_ _

CRITICAL POINTS OF THE t DISTRIBUTION

The first column lists the number of degrees of freedom (ν). The headings of the other columns give probabilities (P) for t to exceed the entry value. Use symmetry for negative t values.

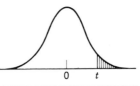

df \ P	.10	.05	.025	.01	.005
1	3.078	6.314	12.706	31.821	63.657
2	1.886	2.920	4.303	6.965	9.925
3	1.638	2.353	3.182	4.541	5.841
4	1.533	2.132	2.776	3.747	4.604
5	1.476	2.015	2.571	(3.365)	4.032
6	1.440	1.943	2.447	3.143	3.707
7	1.415	1.895	2.365	2.998	3.499
8	1.397	1.860	2.306	2.896	3.355
9	1.383	1.833	2.262	2.821	3.250
10	1.372	1.812	2.228	2.764	3.169

48. You want to set a 99% confidence interval for μ on the basis of a sample of 15. Look up the appropriate t_0 in Table V on page 270.

$t_0 = $ _____

_ _

2.977

You will notice that the t table has entries for df of more than 30. If the population is normally distributed but σ is unknown, the t table can be used to obtain a precise confidence interval for μ, more accurate than the approximation obtained when you use s as an estimate of σ with the normal distribution table. The larger the sample, the closer t_0 comes to z_0, so that for large samples there is little practical difference between the two procedures.

49. For each of the following samples how would you go about establishing a 98% confidence interval for μ? Choose the correct answer for each sample.

Sample A

The population from which this sample was drawn has a known standard deviation: $\sigma = 7.85$. The population distribution is known not to be normal.

(a)   Use z

(b)   Use t

(c)   Can't establish a confidence interval

Sample B

The population is assumed to have a normal distribution.

(a)   Use z

(b)   Use t

(c)   Can't establish a confidence interval

Sample C

The standard deviation of the population is unknown. The distribution is known not to be normal.

(a)   Use z
(b)   Use t
(c)   Can't establish a confidence interval

Sample D

The population is known to be normally distributed, with a standard deviation: $\sigma = 8.0$.

(a)   Use z
(b)   Use t
(c)   Can't establish a confidence interval

Sample E

The population is known to be normally distributed.

(a) Use z
(b) Use t
(c) Can't establish a confidence interval

Sample F

On the basis of experience you assume the population is not normally distributed.

(a) Use z
(b) Use t
(c) Can't establish a confidence interval

_ _

Sample A	(a)	Use z	The sample is large
Sample B	(b)	Use t	The sample is small but from a normal population
Sample C	(a)	Use z	The sample is large
Sample D	(b)	Use t	The sample is small but from a normal population
Sample E $\begin{cases}\text{(a)} \\ \text{or} \\ \text{(b)}\end{cases}$	(a) or (b)	Use z Use t	The sample is large and from a normal population*
Sample F	(c)	Can't establish a confidence interval	The sample is small and the population is not normally distributed

*The results will be almost identical, whichever method you use. Theoretically t is more accurate, since you do not know the exact value of σ.

LARGE SAMPLE CONFIDENCE INTERVALS FOR P

Just as you can find confidence intervals for μ by using the sampling distribution of $\bar{x}$, you can also find confidence intervals for P by using the sampling distributions for p; for example, if half a random sample of 30 visitors at a given resort are men ($p = 0.50$), you can state with 99% confidence that between 0.27 and 0.73 of all visitors are male. For small samples special tables based on the binomial probability distributions are available. For large samples the binominal probability distribution becomes approximately normal in shape. Therefore the normal probability table can be used to set a confidence interval for P if the sample is large.

50. Suppose you are trying to estimate the effectiveness of a vaccine. You administer the vaccine to 200 people and then test to determine whether they have acquired an immunity to the disease in question. You find that 185 of 200 have acquired an immunity; that is, $p =$ _____.

- -

0.925 (185/200)

51. You would like to establish a 99% confidence interval for P. First you must make sure the sample is large enough. To establish a confidence interval for P the smaller group in your sample must have at least 10 cases. The smaller group in this case consists of the 15 cases who did not acquire an immunity. Is the sample large enough?

- -

Yes

52. The formula for establishing a large-sample confidence interval for P is

$$p \pm z_0 \frac{\sqrt{pq}}{\sqrt{n}}$$

In this formula p, as you know, is the probability of whatever outcome you are concerned with. The probability of any other outcome is q; for example, if p is the probability of acquiring immunity, q is the probability of not acquiring immunity: $p + q = 1$. If $p = 0.925$, $q =$ _____.

-- -- -- -- -- -- -- -- -- -- -- -- --

0.075

53. Use the formula to establish a confidence interval for p. The z_0 you use is the same z_0 you would use to establish a large-sample confidence interval for μ; P is between _____ and _____ with 99% confidence.

-- -- -- -- -- -- -- -- -- -- -- -- --

0.973 and 0.877

$$p \pm z_0 \frac{\sqrt{pq}}{\sqrt{n}}$$

$$0.925 \pm 2.58 \frac{\sqrt{0.069375}}{\sqrt{200}} = 0.925 \pm 0.048$$

54. In a sample of 100 cases 20 people who receive a fund-raising letter contribute and 80 do not. Predict the proportion P you would expect to contribute if the same letter were sent to the population of 10,000 from which the sample was selected. Use a 95% confidence interval.

-- -- -- -- -- -- -- -- -- -- -- -- -- --

$$p \pm z_0 \frac{\sqrt{pq}}{\sqrt{n}} = 0.20 \pm 1.96 \frac{\sqrt{(0.20)(0.80)}}{\sqrt{100}} = 0.20 \pm 0.08$$

P is between 0.12 and 0.28 with 95% confidence.

REVIEW PROBLEMS

If you have successfully completed this chapter, you can now use sample statistics to estimate population parameters. You can:

● estimate μ, P, and σ;
● establish a confidence interval for μ, using a sample mean and standard deviation;
● establish a confidence interval for P, using data from a large sample.

Now try these review problems. Table I on pp. 253–256 lists any formulas you may need for reference.

1. The following data are a sample selected from a population you assume to be normally distributed. What is your best estimate of μ and σ? Establish a 95% confidence interval for μ.

Data: 3, 4, 5, 6, 6, 7, 7, 7, 8, 8, 8, 9, 9, 10, 11, 12

2. In an appropriately selected sample of 144 members of a fraternal organization, 20% are college graduates. Estimate the percentage you would find if you surveyed all 200,000 members. Establish a 99% confidence interval.

3. You select a random sample of 100 chocolate-covered peanuts from the output of a candy factory. Measuring the thickness of the chocolate coating yields the following statistics for this sample:

$\bar{x} = 0.1$ cm $s = 0.01$ cm

Establish a 95% confidence interval for the parameter μ.

4. State briefly the assumptions involved in establishing a confidence interval using z and t.

Answers

To review a problem, study the frames indicated after the answer.

1. The best estimate of μ is $\bar{x}$.

$$\bar{x} = \frac{\Sigma x}{n} = \frac{120}{16} = 7.50$$

The best estimate of σ is s.

$$s = \sqrt{\frac{\Sigma(x - \bar{x})^2}{n - 1}} = \sqrt{\frac{88}{15}} = \sqrt{5.87} = 2.42$$

For a confidence interval you must use t because the sample is small.

$$t_0 = 2.13 \quad (df = 15)$$

The formula for the confidence interval is

$$\bar{x} \pm t_0 \frac{s}{\sqrt{n}} = 7.5 \pm 2.13 \frac{2.42}{\sqrt{16}}$$
$$= 7.5 \pm 2.13 \, (0.605) = 7.5 \pm 1.29$$

μ is between 6.21 and 8.79 with 95% confidence. *See frames 37 to 49.*

2. This is a large-sample confidence interval for P.
   The formula is

$$p \pm z_0 \frac{\sqrt{pq}}{\sqrt{n}} = 0.20 \pm 2.58 \frac{\sqrt{(0.20)(0.80)}}{\sqrt{144}}$$
$$= 0.20 \pm 2.58 \frac{0.4}{12} = 0.20 \pm 0.085$$

P is between 0.115 and 0.285 with 99% confidence. *See frames 50 to 54.*

3. Because of the large sample size, the appropriate formula is

$$\bar{x} \pm z_0 \frac{s}{\sqrt{n}} = 0.1 \pm 1.96 \frac{0.01}{\sqrt{100}} = 0.1 \pm 0.00196$$

The mean thickness of the chocolate in the factory's output is between 0.098 and 0.102 cm with 95% confidence. *See frames 21 to 36.*

4. Using z to establish a confidence interval assumes that the sample is more than 30. Either σ is known or s is used as an estimate. Using t to establish a confidence interval assumes that the population is normally distributed; σ is never used, always s. *See frames 32 to 36, 37 to 42, and 49.*

CHAPTER FOUR

Hypothesis Testing

In scientific research, we develop theories and then conduct experiments to test them. Since we cannot achieve perfect control in any experiment, some chance variation always results. For example, even the most precise measurement system has some limit to its accuracy, and the laboratory mice bred for biological research to be as similar as possible are still not completely identical. Also, there is a limit to the number of times observations can be repeated, so our measurements are always only a sample of reality.

To generalize from our experimental sample, we use statistical techniques to test how big a part chance plays in the outcome of an experiment. We begin by assuming that our experimental results reflect only the random variations caused by assorted factors beyond our control. This assumption is called the *null hypothesis*. If our experiment is successful and our theory is true, we will be able to reject the null hypothesis by showing that chance variation is not a reasonable explanation for our results.

When you have completed this chapter, you will be able to:

- state a null hypothesis and alternative and establish a critical region for theories about P;
- perform a similar statistical test for theories about μ;
- judge the probability of accepting a theory when it is false or rejecting a theory when it is true.

HYPOTHESIS TESTING—PROPORTIONS

To understand how statistical tests are used in the testing of scientific theories, we will begin by looking at how to use the binomial probability table to test theories about proportions.

1. There is a formal procedure for the statistical testing of theories in scientific research:

    (a) Plan an experiment so that if the results cannot be explained by the chance variation involved in drawing a sample your theory will be confirmed.

    (b) Conduct the experiment and collect sample data.

    (c) Assume that the results are due to chance alone. This assumption is called the null hypothesis.

(d)   Use a theoretical sampling distribution based on the null hypothesis to determine the probability of obtaining sample data like yours by chance alone.

(e)   If the probability of obtaining sample data like yours by chance alone is less than some predetermined small percentage (usually 5% or 1%) the results will be significant. You may reject the null hypothesis and consider your theory confirmed.

In this procedure the null hypothesis assumes that the results of the experiment are due to _____.

_____

chance, the chance variation involved in drawing a sample

2.   If you reject the null hypothesis, your theory (is/is not) confirmed.

_____

is

3.   Study these definitions:

Null hypothesis   assumption that experimental results are due to chance alone

Alternative   your theory (will be confirmed if you reject the null hypothesis)

Significant results   experimental results that are not likely to have occurred by chance alone

The assumption that experimental results are due to chance is called the

_____.

_____

null hypothesis

4.   Your theory is called the _____.

_____

alternative

5.   If the results are significant, you will reject the _____.

_____

null hypothesis

6. Nonsignificant results (will/will not) allow you to reject the null hypothesis.

– –

will not

7. Results that are unlikely to have occurred by chance are called _____.

– –

significant

8. Let us look at an example of this procedure. A researcher is studying the be-havior of fruit flies. He wants to investigate the theory that "you can catch more flies with honey than with vinegar." He establishes a standardized procedure for using both substances as bait and he plans to count the total number of flies caught, n, and the proportion caught with honey, p. His null hypothesis is that there is no real difference between the number of flies that can be caught with honey or vinegar. The differences are due only to chance variations from sample to sample. If this is true, the population parameter $P = $ _____.

– –

0.5. Flies show no real preference in how they get caught. They might as well just flip a coin.

9. The researcher decides to continue the experiment until he catches 15 flies. He can now find a theoretical sampling distribution based on the null hypothesis by looking at a binomial probability table. For $P = 0.5$ and $n = 15$ the distribution is as follows:

14	0.05	0.1	0.2	0.25	0.3	0.4	0.5	0.6	0.7	0.75	0.8	0.9	0.95
15 0	0.463	0.206	0.035	0.013	0.005								
1	0.366	0.343	0.132	0.067	0.031	0.005							
2	0.135	0.267	0.231	0.156	0.092	0.022	0.003						
3	0.031	0.129	0.250	0.225	0.170	0.063	0.014	0.002					
4	0.005	0.043	0.188	0.225	0.219	0.127	0.042	0.007	0.001				
5	0.001	0.010	0.103	0.165	0.206	0.186	0.092	0.024	0.003	0.001			
6		0.002	0.043	0.092	0.147	0.207	0.153	0.061	0.012	0.001	0.001		
7			0.014	0.039	0.081	0.177	0.196	0.118	0.035	0.013	0.003		
8			0.003	0.013	0.035	0.118	0.196	0.177	0.081	0.039	0.014		
9			0.001	0.003	0.012	0.061	0.153	0.207	0.147	0.092	0.043	0.002	
10				0.001	0.003	0.024	0.092	0.186	0.206	0.165	0.103	0.010	0.001
11					0.001	0.007	0.042	0.127	0.219	0.225	0.188	0.043	0.005
12						0.002	0.014	0.063	0.170	0.225	0.250	0.129	0.031
13							0.003	0.022	0.092	0.156	0.231	0.267	0.135
14								0.005	0.031	0.067	0.132	0.343	0.366
15									0.005	0.013	0.035	0.206	0.463

How often would he expect to catch 10 or more flies with honey by chance alone?

0.151 or 15.1% of the time. You must add the frequencies for 10, 11, 12,..., 15. Note: the blank areas in the table represent probabilities so small that they round off to zero.

10. The researcher decides that he will reject the null hypothesis only if his results would occur by chance less than 2% of the time. How many flies will he have to catch with honey to reject the null hypothesis? Use the table in frame 9.

At least 12. The frequencies for 12, 13, 14, 15 add up to 0.017 or 1.7%. Catching 11 or more flies would occur by chance 0.059 or 5.9% of the time, so 12 is the smallest number that will meet the researcher's criterion.

11. The researcher catches 13 flies with honey and 2 with vinegar. Are his results significant?

- -

Yes

12. The set of results that are significant is called the critical region. In this case the critical region is $p \geq \frac{12}{15}$ (p is equal to or greater than $\frac{12}{15}$). Is $\frac{14}{15}$ in the critical region?

- -

Yes

13. Is $\frac{12}{15}$ in the critical region?

- -

Yes

14. For this experiment, $p \geq \frac{12}{15}$ is called the _____.

- -

critical region

15.

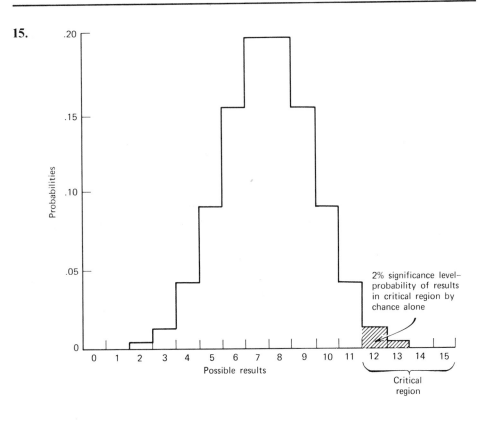

The significance level is the probability that significant results will occur by chance. For this experiment the researcher is using a significance level of ___%.

- -

2%

16. The probability of obtaining results in the critical region by chance is called the

_____.

- -

significance level

17. We can summarize the statistical test of the researcher's hypothesis as follows:

Null hypothesis $P = 0.5$
Alternative $P > 0.5$ (P is greater than 0.5)
Significance level 2%
Critical region $p \geq \frac{12}{15}$ (p is equal to or greater than $\frac{12}{15}$)

If, in fact, you can catch more flies with vinegar than with honey, will the researcher find it out from this statistical test?

— —

No. Of course, he could have started out with the theory that you can catch more flies with vinegar, have considered a fly caught with vinegar a success, and used a similar test.

18. Let us restate the researcher's hypothesis slightly and see how this affects the statistical test. "You will catch a different number of flies with honey than with vinegar." If you can catch flies only with vinegar ($P = 0$), will this theory be confirmed?

— —

Yes

19. If you can catch flies only with honey ($P = 1$), will this theory be confirmed?

— —

Yes

20. What is the appropriate null hypothesis to test this theory?

— —

$P = 0.5$

21. Which of the following is the appropriate alternative for a statistical test of this theory?

(a) $P > 0.5$ (P is greater than 0.5)
(b) $P < 0.5$ (P is less than 0.5)
(c) $P \neq 0.5$ (P is not equal to 0.5)

— —

(c) $P \neq 0.5$

22. Which of the illustrations below shows an appropriate critical region for this statistical test?

Critical
region   (a)

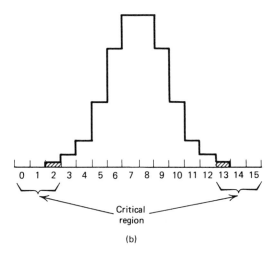

Critical
region
(b)

(b).   In this case either a very high p or a very low p would be consistent with the alternative and would cause you to reject the null hypothesis.

23. To establish the critical region for this statistical test you must consider both ends of the sampling distribution; for example, how often would you expect $p \leq \frac{3}{15}$ (p equal to or less than $\frac{3}{15}$) or $p \geq \frac{12}{15}$ (p equal to or greater than $\frac{12}{15}$)?

0.034 or 3.4% of the time. You must add the frequencies for 0, ..., 3 and 12, ..., 15.

24. If the researcher continues to use a significance level of 2%, what will his critical region be?

p ——————— or p ——————— .

$p \leq \frac{2}{15}$ or $p \geq \frac{13}{15}$

25. Summarize the statistical test for the theory that "you will catch a different number of flies with honey than with vinegar."

Null hypothesis	_____
Alternative	_____
Significance level	_____2%_____
Critical region	_____

Null hypothesis	$P' = 0.5$
Alternative	$P \neq 0.5$
Critical region	$P \leq \frac{2}{15}$ or $P \geq \frac{13}{15}$

26. A caution before we go on. The statistical test of a hypothesis takes into account the chance variations involved in drawing a sample. It does not take into account unplanned influences in the way we conduct the experiment. In a poorly planned experiment we may obtain statistically significant results that are meaningless; for example, suppose that there is a prevailing wind in the direction of the honey at the time we conduct our fruit fly experiments. We catch $\frac{13}{15}$ flies with honey—a statistically significant result. Does this result support the theory that you can catch more flies with honey than with vinegar? Why?

No. You cannot tell whether it was the honey or the direction of the wind that determined the outcome of the experiment.

27. You want to test the theory that a different number of women than men attend sports events. In an appropriately selected sample of 15 attendees you count the proportion of women, p. What is an appropriate null hypothesis for a statistical test of your theory?

_ _

$P = 0.5$

28. What is an appropriate alternative?

_ _

$P \neq 0.5$. You have no firm prediction that women or men are more likely to attend.

29. Use the table in frame 9 to determine a critical region for a significance level of 5%. What is the critical region?

_ _

$p \leq \frac{3}{15}$ or $p \geq \frac{12}{15}$

30. You want to test the theory that more men than women attend sports events. In an appropriately selected sample of 15 attendees you count the proportion of women, p. Summarize an appropriate statistical test for your theory, using a 5% significance level.

Null hypothesis   _____
Alternative   _____
Critical region   _____

_ _

Null hypothesis $P = 0.5$
Alternative $P < 0.5$
Critical region $p \leq \frac{3}{15}$

31. You want to test the theory that more men than women attend sports events. Your research assistant, an attractive young woman, records data on the first 25 people she meets at the stadium. The results are statistically significant. Comment on the meaning of these results.

– –

The experiment was poorly planned. All attendees cannot be assured an equal probability of being selected for the sample. As a result, the data are meaningless as a test of your theory. You might use an experiment of this nature as a test of the theory that an attractive young woman will meet more men than women at sports events.

HYPOTHESIS TESTING—MEANS

So far you have used the hypothesis-testing procedure for theories about the parameter P, but you can use the same kind of procedure for theories about μ.

32. Our fruit fly researcher believes feeding fruit flies a mixture of honey and vinegar will alter their life-span. He knows from long experience that the particular strain of fruit fly he is working with has a mean life-span of 12 days, with a standard deviation of two, when fed their normal diet of apple juice. He plans to raise a sample of 50 fruit flies on honey and vinegar and calculate their mean life-span. What is an appropriate null hypothesis for a statistical test of this theory?

– –

$\mu = 12$

33.   What is the alternative?

$\mu \neq 12$

34.

> **Central Limit Theorem.** For large samples ($n > 30$).
>
> $$\mu_{\bar{x}} = \mu, \qquad \sigma_{\bar{x}} = \frac{\sigma}{\sqrt{n}}$$

The central limit theorem allows him to develop a sampling distribution based on his null hypothesis. The mean of the sampling distribution will be _____.

12

35.   The standard deviation of the sampling distribution will be _____ = 0.28.

$\dfrac{2}{\sqrt{50}}$

36.   The procedure for establishing a critical region in this case is the same as that for establishing a confidence interval for μ. If our fruit fly researcher uses a 1% significance level (equivalent to a 99% confidence interval), what will be the critical region? $z \geq$ _____ or $z \leq$ _____.

$z \geq +2.58$ or $z \leq -2.58$

37. $z_{\bar{x}} = \dfrac{\bar{x} - \mu_{\bar{x}}}{\sigma_{\bar{x}}}$ or $\dfrac{\bar{x} - \mu}{(\sigma/\sqrt{n})}$

If the researcher finds that the mean life-span of his sample of 50 fruit flies is 12.5 days, can he reject the null hypothesis?

No ($z = +1.77$)

38. Summarize the statistical test:

Null hypothesis   _____

Alternative   _____

Critical region   _____

- -

Null hypothesis $\mu = 12$
Alternative $\mu \neq 12$
Critical region $z \geq +2.58$ or $z \leq -2.58$.

39. A company has administered a brief verbal intelligence test to all employees for a number of years. For all 6000 present employees the mean score is 50, with a standard deviation of 10. A personnel researcher wishes to investigate the theory that first-line supervisors score higher on this test than the average employee. To test this theory she selects a random sample of 100 first-line supervisors and computes their mean score on the test. In this problem you are concerned only with one end of the sampling distribution. Summarize an appropriate statistical test to confirm this theory at the 1% level of significance.

Null hypothesis   _____

Alternative   _____

Critical region   _____

- -

Null hypothesis $\mu = 50$
Alternative $\mu > 50$
Critical region $z \geq +2.58$

40. If the mean score of the sample is 53, are the results significant? Refer to the reference formulas in Table I if you wish.

- -

Yes

$$z = \frac{\overline{X} - \mu_{\overline{X}}}{\sigma_{\overline{X}}} = \frac{\overline{X} - \mu}{(\sigma/\sqrt{n})} = \frac{53 - 50}{(10/\sqrt{100})} = 3.0$$

THE PROBABILITY OF ERROR

When you test theories using statistical procedures, there is always a certain possibility that an unusual randomly selected sample will lead you astray. You cannot eliminate the possibility of error, but you can calculate how great the chance of error is and, if necessary, change your experiment to keep it within acceptable limits.

41. A statistical test of a theory never gives absolute certainty. Even when an experiment is planned perfectly, there are two possible types of error:

    Type I   The theory is not true (the null hypothesis is true) but the results are significant by chance.

    Type II   The theory is true (the null hypothesis is false) but the results are not significant.

    As an example, if you use a 5% significance level, you will obtain significant results by chance 5% of the time. This is an example of a Type _____ error.

    _____

    Type I error

42. Type I error occurs when the theory is (true/false) and the results are (significant/not significant).

    _____

    false
    significant

43. If the theory is true but the results are not significant, it is called a Type _____ error.

    _____

    II

44. A Type II error occurs when the theory is (true/false) and the results are (significant/not significant).

    _____

    true
    not significant

45. Label the situations that correspond to Type I and Type II errors:

    (a)   Theory false, results significant
    (b)   Theory true, results significant
    (c)   Theory false, results not significant
    (d)   Theory true, results not significant

_ _

    (a)   Theory false, results significant; Type I error
    (b)   Theory true, results significant
    (c)   Theory false, results not significant
    (d)   Theory true, results not significant; Type II error

46. In 100 similar studies of the difference between two teaching methods 75 studies result in significant differences between the two methods and 25 find no significant difference. If you believe that all the studies were well designed, you will probably consider the 25 studies in which no significant difference was

found to be examples of Type _____ errors.

_ _

    II

47. Seventy-five studies are conducted to determine the effect of a particular drug on the appetite of white rats. The sample sizes vary and several different measures of appetite are used. Four of the studies have results significant at the 5% level. What type of error do you think may be involved here?

_ _

    Type I.   In a large number of studies you would expect a certain number of significant results by chance alone.

48. The significance level you use determines the probability of a Type ___ error.

_ _

    I

49. With a significance level of 1%, the probability of a Type I error is _____.

- -

0.01 (1% of the time)

50. The Greek letter alpha (α) is used as the symbol for the probability of a Type I error. For a statistical test using a significance level of 5% we say $\alpha = 0.05$. For a statistical test using a significance level of 2% we say _____.

- -

$\alpha = 0.02$

51. The Greek letter beta (β) is used as the symbol for the probability of a Type II error. Thus the symbol for the probability of Type I error is ___ and the symbol for the probability of Type II error is ___.

- -

α
β

52. β is the probability of a Type _____ error.
α is the probability of a Type _____ error.

- -

II
I

53. If a researcher's theory is true and β is very large, the chances of proving the theory with a statistical test are (excellent/poor).

- -

poor. He is likely to make a Type II error.

54. Instead of referring to β, researchers sometimes use the term "power." Power is $(1 - \beta)$. A statistical test with a low β has a (high/low) power.

- -

high

55. A statistical test for which there is a high probability of a Type II error has (high/low) power.

- -

low

56. Would you say that power in a statistical test is a good thing?

- -

yes. You have a better chance of confirming your theories if they are correct.

57. If you make an assumption about the true state of affairs, you can determine the power of a statistical test; for example, let us consider the test at the start of this chapter for the theory that "You can catch more flies with honey than with vinegar." As you recall, the experiment consisted of catching 15 flies and counting the number caught with honey as opposed to vinegar. A significance level of 2% was chosen ($\alpha = 0.02$).

Null hypothesis	$P = 0.5$
Alternative	$P > 0.5$
Critical region	$P \geq \frac{12}{15}$

To reject the null hypothesis we must catch at least _____ of the 15 flies with honey.

- -

12

58. Let us assume that if we could test all flies 80% would prefer honey ($P = 0.8$). We can now use the binomial probability table to give us a sampling distribution for samples of 15 from this population. The distribution is as shown below.

14

15		0.05	0.1	0.2	0.25	0.3	0.4	0.5	0.6	0.7	0.75	0.8	0.9	0.95
	0	0.463	0.206	0.035	0.013	0.005								
	1	0.366	0.343	0.132	0.067	0.031	0.005							
	2	0.135	0.267	0.231	0.156	0.092	0.022	0.003						
	3	0.031	0.129	0.250	0.225	0.170	0.063	0.014	0.002					
	4	0.005	0.043	0.188	0.225	0.219	0.127	0.042	0.007	0.001				
	5	0.001	0.010	0.103	0.165	0.206	0.186	0.092	0.024	0.003	0.001			
	6		0.002	0.043	0.092	0.147	0.207	0.153	0.061	0.012	0.003	0.001		
	7			0.014	0.039	0.081	0.177	0.196	0.118	0.035	0.013	0.003		
	8			0.003	0.013	0.035	0.118	0.196	0.177	0.081	0.039	0.014		
	9			0.001	0.003	0.012	0.061	0.153	0.207	0.147	0.092	0.043	0.002	
	10				0.001	0.003	0.024	0.092	0.186	0.206	0.165	0.103	0.010	0.001
	11					0.001	0.007	0.042	0.127	0.219	0.225	0.188	0.043	0.005
	12						0.002	0.014	0.063	0.170	0.225	0.250	0.129	0.031
	13							0.003	0.022	0.092	0.156	0.231	0.267	0.135
	14								0.005	0.031	0.067	0.132	0.343	0.366
	15									0.005	0.013	0.035	0.206	0.463

What is the probability of catching 12 or more flies with honey from this population?

- -

0.648, about 65% of the time

59. We can say that the power of our test against the alternative $P = 0.8$ is 0.648. The chance of rejecting the null hypothesis in this case is 0.648. Therefore the chance of accepting it incorrectly is _____ .

- -

$1 - 0.648 = 0.352$

60. If $P = 0.8$, how often will this statistical test result in a Type II error? $\beta = $ _____.

$\beta = 0.352$

61. What is the power of this test against the alternative $P = 0.9$? You will use the same reasoning but now you must use the sampling distribution for $P = 0.9$.

The power of the test against the alternative $P = 0.9$ is 0.945.

You can make similar computations for statistical tests about the mean and for other statistical tests covered in this book. The details of the computations are not presented, but the logic of the process remains the same as in the simple case we have considered. You must make enough assumptions about the alternative to allow you to construct a theoretical sampling distribution for the alternative; for example, for a test about the mean you must assume values of μ and σ. Then compare this new sampling distribution with the critical region established for your statistical test to determine how often you would obtain significant results under the new assumption.

62. Three factors that influence the probability of a Type II error are significance level, size of the sample, and variability of the population. A large α results in a relatively smaller β; for example, an α of 0.05 will cause (more/less) risk of a Type II error than an α of 0.01.

----

less. You are more likely to reject the null hypothesis even if it is true.

63. An α of 0.05 will cause (more/less) risk of a type I error than an α of 0.01.

----

more

64. A large sample will result in relatively less risk of a type II error; β against a given alternative will be greater with an n of (35/100).

_ _

35

65. If the populations have relatively large standard deviations, β will be relatively greater. In which situation is the risk of a Type II error greater:

(a) σ appears to be about 2.
(b) σ appears to be about 10.

_ _

(b)

66. All other things being equal, the power of a statistical test will be greater with an α of (0.05/0.01).

_ _

0.05

67. All other things being equal, the power of a statistical test will be greater with an n of (30/75).

_ _

75

68. All other things being equal, the power of a statistical test will be greater if the standard deviations of the data appear to be in the range of (5/25).

_ _

5

69. You are screening anticancer drugs for possible beneficial effects. You know that none of them is harmful but you are uncertain whether they do any good. If your results are significant for a given drug, it will be the subject of further intensive investigation. If your results are not significant, the investigation will be dropped. Which type of error would you be more concerned about avoiding (Type I/Type II)?

- -

Type II. This error would result in discarding a useful drug. A Type I error would result only in wasted effort in further investigation.

70. In this situation you would probably choose an α of (0.05/0.001).

- -

0.05

71. You are investigating a theory that runs completely counter to all accepted theory in your field. In some preliminary studies with small samples you obtained results significant at the 5% level. You have now designed an experiment with a very large sample. You have refined your measurement techniques so that you believe the standard deviations of your data will be relatively small. You will probably choose an α of (0.05/0.01).

- -

0.01. In this case you are most concerned about avoiding a Type I error.

REVIEW PROBLEMS

If you have successfully completed this chapter you can now set up a formal statistical test of a theory. You can:

- state a null hypothesis and an alternative;
- establish a critical region for p to test a hypothesis about P;
- establish a critical region for $\bar{x}$ to test a hypothesis about μ;
- comment on the practical importance of Type I and Type II error in given situations.

Now try these review problems. Table I on pp. 253–256 lists any formulas you may need for reference.

1. "We conducted an experiment based on our theory and the results were statistically significant at the 1% level. Therefore our theory is conclusively proved." Comment on this assertion.

2. "For this proposed experiment the probability of a Type II error appears to be so high that it is not worth the effort to collect the data." What does this mean?

3. A clairvoyant claims he can predict whether a coin will land heads or tails. In an experiment that tested his claim, a coin was tossed 10 times and he predicted correctly eight times. Outline an appropriate statistical test. Are these results significant at the 5% level?

4. How often would you expect to obtain eight or more correct predictions by chance in the experiment in problem 3?

5. The mean height of adult males in a particular ancient culture was 5ft 2 in. with a standard deviation of 2in., as determined by measurement of a large number of skeletons found in burial sites. A new site has been found that differs somewhat from the others. The discoverer theorizes that the skeletons at this site are from a different racial stock, with a different mean height. Outline an appropriate statistical test with a 1% significance level.

Answers

To review a problem, study the frames indicated after the answer.

1. The statistical test indicates that the results are not likely to have occurred as a result of the chance variations involved in sampling. You would, however, have to examine the design of the experiment carefully to be sure that the experimenter's theory is the only reasonable explanation of the results. Even if the experimental design were perfect, there would be a 1% chance that the theory would not be correct. *See frames 26, 31, and 41 to 71.*

2. Even if the theory were true, you are not likely to obtain results that would allow you to reject the null hypothesis. You are unlikely to obtain significant results. *See frames 41 to 71.*

3. Let us call a correct prediction a success. On the average one should predict correctly half the time just by chance.

Null hypothesis	$P = 0.5$
Alternative	$P > 0.5$
Significance level	$\alpha = 0.05$
Critical region	$p \geq \frac{9}{10}$

    Since $p = \frac{8}{10}$, the results are not significant. *See frames 1 to 31.*

4. The probabilities in the binomial probability table are

8	0.044
9	0.010
10	0.001
	0.055

    You would expect eight or more successes 5.5% of the time. *See Chapter 2, frames 52 to 71.*

5. | | |
    |---|---|
    | Null hypothesis | $\mu = 5\text{ft } 2\text{in.}$ |
    | Alternative | $\mu \neq 5\text{ft } 2 \text{ in.}$ |
    | Significance level | $\alpha = 0.01$ |
    | Critical region | $z \leq -2.58$ or $z \geq +2.58$ |

    See frames 32 to 40.

CHAPTER FIVE
Differences Between Means

This chapter is about statistical tests we can use to investigate the difference between the means of two sets of observations. You will learn about two situations. One is the case where your observations come in pairs. For example, you may have "before" and "after" observations, or you may have observations of situations that are carefully matched except for one factor whose effect you are investigating. The other case is where you have two samples selected at random.

In each of these cases, we can use a statistical test to see if the differences can be explained by chance variations or if we may reject this null hypothesis and say that the differences support an experimental hypothesis.

When you have completed this chapter, you will be able to:

- test the significance of a set of difference scores;
- test the significance of the difference between the means of two independent samples;
- recognize when you have independent samples and when difference scores, and apply the appropriate test procedure to each.

DIFFERENCE SCORES

Difference scores are the result of taking two related observations, for example "twin A" and "twin B," "before" and "after." When dealing with paired data of this sort, it is important to remember that the statistical analysis is based on the differences between the paired scores.

1. One common type of experimental study is the "before" and "after" study. Let us consider an example. We wish to test the theory that people will lose weight if they stick to a diet of grapefruit and whole wheat toast. To test this theory we select a sample of 49 people, weigh them at the start of our study, and weigh them again after a period of dieting. For each person we subtract the "before" weight from the "after" weight to obtain a difference score; for example, a person who weighs 125 lb at the start of the study and

    118 lb at the end of the study will have a difference score of _____.

    - - - - - - - - - - - - - - - - - -

    -7

2. A person who weighs 140 lb at the start of the study and (because he loves whole wheat toast) 150 lb at the end of the study will have a difference score of _____.

_ _

+ 10

3. If on the average the diet makes no difference, the mean difference score for the population is _____.

_ _

zero

4. If, on the average, people lose weight on this diet, the mean difference score is _____.

_ _

less than zero (negative)

5. What is an appropriate null hypothesis for a statistical test of our theory that people will lose weight on this diet?

_ _

The mean difference score for the population is zero.

6. What is an appropriate alternative?

_ _

The mean difference score for the population is less than zero.

7. If we can construct a theoretical sampling distribution based on the null hypothesis, we can make a statistical test of our theory. Let us consider what the
population of difference scores would be like if the diet had no effect. We have
said that the mean difference score would be zero. Does this mean that the
difference score of every person who adheres to the diet will be zero?

– – – – – – – – – – – – – – – – – – – –

No. Some will lose weight for reasons that have nothing to do with the diet.
Others will gain weight. Among a sufficiently large number of dieters these
variations should tend to cancel out to produce a mean of zero.

8. In the preceding problems in which you tested a hypothesis about the mean
your null hypothesis was based on a population whose mean and standard
deviation were known. In this case the null hypothesis is based on a theoretical
analysis. We can say that if the diet has no effect the mean of a population of
difference scores should be zero, but we have no basis in theory for saying what
the standard deviation of this population should be. Instead, we must use
the sample to provide us with an estimate of σ. What statistic will we use?

– –

s

9. As you recall, $\sigma_{\bar{x}} = \sigma/\sqrt{n}$.
Since we do not know σ in this case, our best estimate is $\sigma_{\bar{x}} =$ _____.

– –

$$\frac{s}{\sqrt{n}}$$

10. With this estimate of $\sigma_{\bar{x}}$, we can compute a z score and test our hypothesis.

Null hypothesis $\mu = 0$
Alternative $\mu < 0$
Significance level $\alpha = 0.05$
Critical region $z \leq -1.65$

Our sample provides the following statistics:

$n = 49$
$\bar{x} = -5.0$
$s = 3.5$

Compute $z = \dfrac{\bar{x} - \mu}{\sigma_{\bar{x}}} = \dfrac{\bar{x} - \mu}{s/\sqrt{n}} = $ _____.

Are the results statistically significant?

- -

$$z = \frac{-5.0 - 0}{3.5/\sqrt{49}} = \frac{-5.0}{3.5/7} = \frac{-5.0}{0.5} = -10.0$$

The results are significant.

11. In this example we used s as an estimate of σ. This estimate is good only if the sample size is _____.

- -

larger than 30

12. Let's review this hypothesis-testing procedure. According to the null hypothesis in this case,

$\mu = $ _____

$\sigma = $ _____

- -

$\mu = 0$
$\sigma = 3.5$

13.   Using a significance level of $\alpha = 0.05$, we established a critical region of $z \leq -1.65$. If the null hypothesis is true and we repeat the same experiment a great number of times, how often would we expect to obtain a sample mean with a z score in the critical region?

- -

5% of the time

14.   If the alternative $\mu < 0$ were true, we should expect to obtain a sample mean with a z score in the critical region (more/less) often.

- -

More

A different approach to the same problem would be to use $\overline{x}$ and s to establish a confidence interval for μ. The formula for the confidence interval is

$$\overline{x} \pm z_0 \frac{s}{\sqrt{n}} = -5 \pm 1.65 \frac{3.5}{\sqrt{49}} = -5 \pm 0.825$$

μ is between -5.825 and -4.175 with 90% confidence.

Since cases in which μ is even smaller than -5.825 also confirm the theory, you can say that $\mu \leq -4.175$ with 95% confidence. Clearly $\mu < 0$ with at least 95% confidence. An $\overline{x}$ that is exactly on the boundary of the critical region ($z = -1.65$) will give exactly 95% confidence that $\mu < 0$.

15.   For samples larger than 30 the method we have used is valid. For smaller samples we can use the same method with the t distribution, provided we can assume that the population of difference scores is approximately normally distributed; for example, a psychologist believes that a particular type of memory training will influence the ability of children to remember nonsense syllables. He tests a class of 25 children with a list of nonsense syllables before training and with an equivalent list after training and he finds the difference score for each child. Before he can use a t test on his results, he must check the assumption that the difference scores are_____.

- -

approximately normally distributed.

The mathematical derivation of the t tables assumes that the population is normally distributed. In practice a t test will usually lead to the correct conclusion even when the distribution is fairly different from a normal distribution, provided it is not strongly skewed.

16. If he believes that the distribution of the difference scores is likely to be very different from a normal distribution, what can he do in planning his experiment to ensure that he will be able to make a statistical test of his results?

- -

Use a larger sample

17. Let us assume that the difference scores tend to be normally distributed in this example. Outline an appropriate statistical test below. Use a significance level of 1% and consider both ends of the sampling distribution. Use the t table on page 270 to establish the critical region and remember that df $= n - 1$.

Null hypothesis   _____

Alternative   _____

Significance level   _____

Critical region   _____

- -

Null hypothesis $\mu = 0$
Alternative $\mu \neq 0$
Significance level $\alpha = 0.01$
Critical region $t \geq +2.80$ or $t \leq -2.80$

18. Reference formula

$$t = \frac{\bar{x} - \mu}{s/\sqrt{n}}$$

The experimenter collects the following data:

The mean difference between pretest and posttest is -5 points. The standard deviation of the sample of difference scores is 20 points.

Are the results significant?

- -

$$No \left(t = \frac{-5}{20/\sqrt{25}} = \frac{-5}{4} = -1.25 \right)$$

19. Here is the computer printout of another statistical test:

PAIRED SAMPLES T TEST

VARIABLE	N	MEAN	SD	MEAN DIFF	SD	T VALUE	2-TAIL PROB
PRE		164.3	21.5409				
	10			3.3	0.9434	3.49799	.006
POST		161.0	21.0079				

What is the mean difference between PRE and POST? _____
What is the probability of such a result by chance? _____

- -

3.3
p = .006

20. Difference scores need not always be before and after scores; for example, consider the following situation:

You are studying the effect of environment on intelligence test scores. You are fortunate enough to locate five sets of identical twins one of whom was raised in an institution and the other adopted and raised in a family setting. You test all the twins and obtain the following scores:

Pair No.	Raised in family	Institutionalized	Difference
1	105	95	-10
2	95	83	-12
3	103	103	0
4	98	96	-2
5	103	97	-6

Can you reasonably assume that these difference scores are normally distributed? There is no clear evidence that they are *not* normally distributed and most researchers would accept the use of a *t* test. You should remember, however, that you are making this assumption on very limited evidence. Outline an appropriate statistical test for the theory that the difference in environment makes a difference in the intelligence test scores of these identical twins. Use a significance level of 5%. The *t* table is on page 270.

Null hypothesis　　　_____

Alternative　　　_____

Significance level　　_____

Critical region　　_____

- -

Null hypothesis	$\mu = 0$
Alternative	$\mu \neq 0$
Significance level	$\alpha = 0.05$
Critical region	$t \geq +2.78$ or $t \leq -2.78$

21. $\bar{x} = \dfrac{\Sigma x}{n}$ $\qquad s = \sqrt{\dfrac{\Sigma(x - \bar{x})^2}{n - 1}}$

To compute t you will need to know the mean and standard deviation of the sample. Using the reference formulas above, compute $\bar{x}$ and s. (You will find a square root table on page 257.)

$\bar{x} = $ _____

$s = $ _____

- -

$\bar{x} = \dfrac{-30}{5} = -6.0$

$s = \sqrt{\dfrac{104}{4}} = \sqrt{26} = 5.1$

22. Complete the t test by computing t. Is the difference between the twins statistically significant?

- -

$t = \dfrac{\bar{x} - \mu}{s/\sqrt{n}} = \dfrac{-6.0 - 0}{5.1/\sqrt{5}} = \dfrac{-6}{2.3} = 2.61$

The difference is not statistically significant. Your statistical test does not support the theory that environment affects intelligence test scores.

23. In this experiment the result is not significant, but it is extreme enough that it would not occur by chance as often as 10% of the time. The sample size is small. What would you say about the chance of a Type I or Type II error in this case?

— —

The chance of a Type II error is substantial. This statistical test may well have led us to reject a true theory.

THE DIFFERENCE BETWEEN TWO INDEPENDENT SAMPLE MEANS

The statistical analysis appropriate for comparing two independent samples is not the same as the procedure you have just learned for testing difference scores.

24. A common type of experimental study involves the comparison of an experimental sample and a control sample; for example, an experimenter interested in the effect of a plant hormone on the growth of beanstalks might treat alternate rows of beans with the hormone, leaving half the plants untreated. She will measure the height of each beanstalk and then use a statistical test to determine if the difference in mean height between the two samples of beanstalks is significant. This study does not involve the kind of individual-by-individual matching that was used in the difference score studies. Selecting one bean for the treated group does not give you any information about what beans will be chosen for the other group.
Does selecting one dieter's "before" weight tell you anything about the composition of the "after" sample?

— —

Yes. The "after" sample will include the same person.

25. In which problem would you say the two samples are independent?

(a) The bean problem
(b) The dieting problem

— —

(a) The bean problem

26. For the bean problem there is a statistical test we can use to test the null hypothesis that two independent samples come from populations with the same mean. Let μ_1 be the mean height of the treated plants and μ_2, the mean height of the untreated. We would state the null hypothesis this way:

Null hypothesis $\mu_1 = \mu_2$

How would you state the alternative?

- -

$\mu_1 \neq \mu_2$

27. If the experimenter has a theory that the plant hormone will stunt the growth of the beanstalks, how would you state the alternative? Remember that μ_1 is the treated plants.

- -

$\mu_1 < \mu_2$

28. If both samples are large, we can compute a z score for the difference between means. If the samples are small, we can use the same techniques to compute a t score, provided certain assumptions are true. We will not explain the mathematical derivation of the formulas you will use. The end result is a z score or t score which you can compare with the appropriate table. Let us start with the case of large samples. The experimenter has a theory that predicts that the hormone will stunt the growth of beans and she will consider only one end of the sampling distribution. She will use a significance level of 0.01. Her samples contain 100 beanstalks each. Outline an appropriate statistical test.

Null hypothesis   _____

Alternative   _____

Significance level   _____

Critical region   _____

- -

Null hypothesis $\mu_1 = \mu_2$
Alternative $\mu_1 < \mu_2$
Significance level $\alpha = 0.01$
Critical region $z \leq -2.33$

29. The formula used for computing a z score for the difference between two means is

$$z = \frac{\overline{x}_1 - \overline{x}_2}{\sqrt{s_1^2/n_1 + s_2^2/n_2}}$$

For the following data compute a z score:

Sample 1 (treated)	Sample 2 (untreated)
$n_1 = 100$	$n_2 = 100$
$\overline{x}_1 = 27$ in.	$\overline{x}_2 = 29$ in.
$s_1 = 5$ in.	$s_2 = 4$ in.

- -

$z = 3.13$

$$z = \frac{27 - 29}{\sqrt{5^2/100 + 4^2/100}} = \frac{-2}{\sqrt{25/100 + 16/100}} = \frac{-2}{\sqrt{0.4}} = -3.13$$

The experimenter's theory is supported by the statistical test. Treated beanstalks are significantly shorter.

30.   Reference formula.

$$z = \frac{\overline{x}_1 - \overline{x}_2}{\sqrt{s_1^2/n_1 + s_2^2/n_2}}$$

A manufacturer suspects a difference in the quality of the spare parts he receives from two suppliers. He obtains the following data on the service life of random samples of parts from the two suppliers:

Supplier A	Supplier B
$n_1 = 50$	$n_2 = 100$
$\overline{x}_1 = 150$	$\overline{x}_2 = 153$
$s_1 = 10$	$s_2 = 5$

Outline an appropriate statistical test, using the 1% significance level, and compute z. Is the difference between the two samples statistically significant?

- -

Null hypothesis	$\mu_1 = \mu_2$
Alternative	$\mu_1 \neq \mu_2$
Significance level	$\alpha = 0.01$.
Critical region	$z \leq + 2.58$ or $z \geq - 2.58$

$z = 2.0$ (not significant).

$$z = \frac{150 - 153}{\sqrt{10^2/50 + 5^2/100}} = \frac{-3}{\sqrt{2 + 0.25}} = \frac{-3}{1.5} = -2.0$$

The two suppliers' parts are not significantly different.

31. To use this type of test with small samples two assumptions must be made:

(a)   The two populations involved are normally distributed.

(b)   The populations have equal standard deviation; that is, $\sigma_1 = \sigma_2$.

Since the populations are unknown, you can never be certain if these assumptions are valid, but you can make some judgments about their reasonableness by looking at the data; for example, consider the following cases:

Case 1. You are testing for a significant difference between the following samples:

Sample 1   1, 4, 6, 8, 10, 11, 12, 14, 16, 18
$\bar{x} = 10, s = 5.2$

Sample 2   2, 5, 6, 8, 11, 13, 14, 15, 17, 20
$\bar{x} = 11.1, s = 5.7$

Case 2. You are testing for a significant difference between the following samples:

Sample 1   1, 4, 6, 8, 10, 11, 12, 14, 16, 18
$\bar{x} = 10, s = 5.2$

Sample 2   1, 1, 2, 2, 2, 3, 5, 10, 24, 38
$\bar{x} = 8.8, s = 12.4$

Which case better matches assumption 1 (case 1/case 2)?

_____

case 1. Sample 2 in case 2 is skewed.

32.   Which case better matches assumption 2 (case 1/case 2)?

_____

case 1. The values of s for the two samples in case 2 are quite different; one value is more than twice as large as the other.

33.   Using t to test the significance of the difference between the means of two independent small samples involves two assumptions:

(a)   Both distributions are _____.

(b)   The population standard deviations are _____.

_____

normal
equal

In practice, this t *test will usually lead to the correct conclusion provided the two samples are equal in size and neither of the distributions is strongly skewed.*

34. The formula used for computing a t score for the difference between two means is

$$t = \frac{\bar{x}_1 - \bar{x}_2}{\sqrt{s^2/n_1 + s^2/n_2}}$$

In this formula s^2 is a pooled estimate of the population variance (σ^2), based on the s of both samples combined. The formula for computing s^2 is

$$s^2 = \frac{(n_1 - 1)\, s_1^2 + (n_2 - 1)\, s_2^2}{n_1 + n_2 - 2}$$

For the following data compute t score:

Sample 1	Sample 2
$n = 10.0$	$n = 10.0$
$x = 10.0$	$x = 11.1$
$s = 5.2$	$s = 5.7$

- -

$$s^2 = \frac{9(5.2)^2 + 9(5.7)^2}{10 + 10 - 2} = \frac{9(5.2^2 + 5.7^2)}{18} = \frac{27.04 + 32.49}{2} = \frac{59.53}{2} = 29.765$$

$$t = \frac{10.0 - 11.1}{\sqrt{29.765/10 + 29.765/10}} = \frac{-1.1}{\sqrt{59.53/10}} = \frac{-1.1}{\sqrt{5.95}} = \frac{-1.1}{2.44} = -0.45$$

35. As you recall, when using the t table you must take into account the "degrees of freedom." When you use only one sample, df $= n - 1$, but when you have two samples you must add the degrees of freedom for each sample. Then df $= (n_1 - 1) + (n_2 - 1)$ or, expressed in a simpler form, df $= n_1 + n_2 - 2$. What are the degrees of freedom for the problem in frame 33?

- -

18 (10 + 10 − 2)

36. For the following statistical test, establish a critical region for the problem in frame 33 (df = 18).

Null hypothesis	$\mu_1 = \mu_2$
Alternative	$\mu_1 \neq \mu_2$
Significance level	$\alpha = 0.05$

Critical region _____

- -

t ≥ +2.10 or $t \le -2.10$

37. When you use a computer to calculate a *t* test on the difference between two sample means, the result will look something like this:

T-TEST

	N	MEAN	SD	T VALUE	DF	2-TAIL PROB
GROUP 1	25	83.84	13.31			
				3.50	36	0.00
GROUP 2	13	67.31	14.77			

Look at the printout. What is the value for degrees of freedom? _____ Is it correct? (Yes/No) Are the results of the test significant at the 5% level? (Yes/No)

_____

df = 36
Yes, it is correct.
Yes, the results are significant.

CHOOSING THE RIGHT TEST

Using an inappropriate statistical test will give misleading results. To select an appropriate statistical test you must consider two questions:

(a) What are your null hypothesis and alternative?
(b) What assumptions and limitations apply to the statistical tests you might use?

38. Let us look at some examples. A reading test is supplied by a national publisher. The test is accompanied by a manual which describes the results of the administration of the test to some 4000 tenth-grade students in various parts of the country. According to this manual, the distribution of scores on the test was approximately normal, with a mean of 30, and a standard deviation of 5. You wish to know if the performance of students in your city differs significantly from that of the 4000 students described in the manual, so you select a random sample of 25 tenth-grade students and test them. What are your null hypothesis and alternative for the statistical test?

Null hypothesis _____

Alternative _____

_____

Null hypothesis $\mu = 30$
Alternative $\mu \neq 30$

39. The form of your null hypothesis tells you what statistical tests are available. Which of the following corresponds to your null hypothesis in this case?

(a) $\mu = C$, where C is a known constant that you could state before the experiment; for example, $\mu = 0$ or $\mu = 10$.

(b) $\mu_1 = \mu_2$, where μ_1 and μ_2 are both estimated on the basis of samples.

- -

(a), $\mu = C$

40. Look at the summary of formulas on page 253. What two formulas are listed for testing the hypothesis $\mu = C$?

- -

$$z = \frac{\overline{x} - C}{s/\sqrt{n}}$$
$$t = \frac{\overline{x} - C}{s/\sqrt{n}}$$

41. The form of your alternative tells you whether your test is "one-tailed" or "two-tailed." A two-tailed test is a test in which you take into account both ends of the sampling distribution when you set your critical region. It corresponds to an alternative such as $\mu \neq C$. A one-tailed test is concerned with only one end of the sampling distribution. It corresponds to an alternative such as $\mu \geq C$ or $\mu \leq C$. Consider your alternative in this problem. What type of test will you use?

- -

Two-tailed. You will reject the null hypothesis if the sample mean is either unusually low or unusually high.

42. Now ask yourself the second question: "What assumptions and limitations apply to the statistical tests you might use"? Use of a z score involves an assumption about sample size. What is it?

_ _

The sample size is greater than 30.

43. Using a t score involves an assumption about the distribution of your unknown population. What is it?

_ _

The population is approximately normally distributed.

44. The sample size in this problem is 25. Suppose you test the 25 students and find that their scores appear to be roughly normally distributed, with $\overline{x} = 26$ and $s = 6$. What will you use for a statistical test?

(a) a z score
(b) a t score
(c) neither

_ _

(b) a t score

45. Using a 1% significance level, complete the statistical test:

Null hypothesis $\qquad$ $\mu = 30$
Alternative $\qquad$ $\mu \neq 30$
Significance level $\qquad$ $\alpha = 0.01$

Critical region $\qquad$ $t \geq$ _____ or $t \leq$ _____

$t =$ _____. The results are (significant/not significant).

- -

Critical region; $t \geq +2.797$ or $t \leq -2.797$ (df $= 24$)

$$t = \frac{26 - 30}{6/\sqrt{25}} = \frac{-4}{6/5} = \frac{-20}{6} = -3.33$$

Significant

Your sample does differ significantly from the sample described in the manual.

46. Using the same reading test with a sample of 25 students from a different city, you find that their scores are bimodally distributed, as indicated in the illustration. The city has a large minority of students who have learned English as a second language.

Again, you wish to know if the mean performance, of all students in this city differs significantly from that of the 4000 students described in the test manual. What will you use for a statistical test in this case?

(a)   a *z* score
(b)   a *t* score
(c)   neither

_ _

(c) neither. In this case you cannot assume that the sample comes from a normally distributed population, and the sample is too small to use a *z* score. One solution would be to obtain a larger sample so that a *z* score could be used. Perhaps a more appropriate solution would be to treat the students who learned English as a second language as a separate population and draw two separate samples.

47. All tenth-grade students in your city have been tested. This population consists of 3000 tenth-graders. Their reading scores are distributed as shown below. A group of 100 from this population has been selected for a special

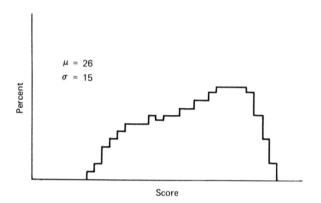

enrichment program and you suspect that the selection procedures favored the better readers. To check your theory you obtain the reading scores of the 100 students selected for the special program. What are the null hypothesis and alternative?

‒ ‒

Null hypothesis $\mu = 26$
Alternative $\mu > 26$

48. Which of the following corresponds to your null hypothesis in this case?

(a) $\mu = C$
(b) $\mu_1 = \mu_2$

‒ ‒

(a) $\mu = C$

49. The form of the null hypothesis indicates that statistical tests are available with *z* scores or *t* scores. The form of the alternative indicates that the test will be (one-tailed/two-tailed).

_ _

one-tailed

50. What assumptions and limitations apply to the available tests? For a *z* score? For a *t* score?

_ _

For a *z* score *n* must be greater than 30.
For a *t* score the sample is assumed to have come from a normally distributed population.

51. Considering these assumptions, what will you use for a statistical test in this case?

(a)   a *z* score
(b)   a *t* score
(c)   neither

_ _

(a) a *z* score. The sample is large enough to use *z*.

Although the population is not normally distributed, most researchers would probably accept conclusions based on a t *score for a smaller sample in this case.*

52. Theoretical analysis leads a researcher to predict that the trout in one lake (call it Deep Lake) will have higher concentrations of chlorinated hydrocarbons in their livers than the trout in another lake (Blue Lake). He catches 10 fish from each lake. What are the null hypothesis and alternative?

––––––––––––––––––––––––––

Null hypothesis $\mu_1 = \mu_2$
Alternative $\mu_1 > \mu_2$

53. Look at the summary of formulas on page 253. What formulas are listed for testing this hypothesis?

––––––––––––––––––––––––––

$$z = \frac{\overline{x}_1 - \overline{x}_2}{\sqrt{s_1^2/n_1 + s_2^2/n_2}}, \qquad t = \frac{\overline{x}_1 - \overline{x}_2}{\sqrt{s^2/n_1 + s^2/n_2}}$$

54. The statistical test will be (one-tailed/two-tailed)?

––––––––––––––––––––––––––

one-tailed

55. What assumptions apply to the z score test in this case?

– –

Both samples are larger than 30.

56. What assumptions apply to the t score test in this case?

– –

The populations are normally distributed *and* have equal standard deviations.

57. The two samples appear to be more or less normally distributed. The following amounts of chlorinated hydrocarbons are detected:

Sample 1 (Deep Lake)	Sample 2 (Blue Lake)
$\overline{x} = 8$	$\overline{x} = 4$
$s = 3$	$s = 2$
$n = 10$	$n = 10$

Review the problem as outlined in frame 50 and the following frames and complete an appropriate statistical test. Do these results confirm the researcher's theory? (Use the 1% significance level.)

– –

Null hypothesis $\qquad \mu_1 = \mu_2$
Alternative $\qquad\qquad \mu_1 > \mu_2$
Significance level $\qquad \alpha = 0.01$
Critical region $\qquad t \geq +2.55$ (df = 18)

$$s^2 = \frac{9(2)^2 + 9(3)^2}{18} = 6.5 \qquad t = \frac{8 - 4}{\sqrt{1.3}} = 3.51$$

The researcher's theory is confirmed. The t test is used because the samples are small and s_1 and s_2 are not drastically different.

58. It is important to distinguish between independent samples and samples that are actually paired data and should be used to compute difference scores; for example, a statistically minded shopper wants to compare the prices of the two supermarkets in his neighborhood. From a list of all types of canned and frozen goods that are available he selects at random 43 different items. He then determines the price of each of these items at each store, so that he has a list of 43 prices for each of the two stores. Do his two lists of prices constitute independent samples or should he compute difference scores?

_ _ _ _ _ _ _ _ _ _ _ _ _ _ _ _ _ _ _ _

He should compute difference scores, since he priced the same items at each store.

59. What are an appropriate null hypothesis and alternative for a statistical test?

_ _

Null hypothesis $\qquad \mu = 0$
Alternative $\qquad\qquad \mu \neq 0$

60. What formula should he use for a statistical test?

_ _ _ _ _ _ _ _ _ _ _ _ _ _ _ _ _ _ _ _

$$z = \frac{\overline{x} - C}{s/\sqrt{n}}$$

61. An economist is investigating price differences between suburban and central city stores. He determines the total price of a standard shopping list at a random sample of 18 suburban stores and 22 central city stores, so that he has 18 shopping-list totals from suburban stores and 22 shopping-list totals from central city stores. Has he two independent samples?

Yes. Selecting a given suburban store has no influence on what central city stores will be selected.

62. Outline an appropriate statistical test.

Null hypothesis   _____

Alternative   _____

Significance level   ____ $= 0.05$ _____

Critical region   _____

Formula for test   _____

Null hypothesis $\mu_1 = \mu_2$
Alternative $\mu_1 \neq \mu_2$
Significance level $\alpha = 0.05$
Critical region $t \geq +2.02$ or $t \leq -2.02$ (df $= 38$)

Formula for test $t = \dfrac{\bar{x}_1 - \bar{x}_2}{\sqrt{s^2/n_1 + s^2/n_2}}$

$s^2 = \dfrac{(n_1 - 1)s_1^2 + (n_2 - 1)s_2^2}{n_1 + n_2 - 2}$

A last caution about one-tailed versus two-tailed tests. A one-tailed test is appropriate only when you have clearly decided in advance of the experiment that results in the opposite direction from your alternative are of no interest whatsoever. You cannot look at the data and then decide to use a one-tailed test. For this reason many researchers always use two-tailed tests even when they have formed a tentative conclusion about the direction of the results. When in doubt, use a two-tailed test.

REVIEW PROBLEMS

If you have successfully completed this chapter, you can now set up formal statistical tests of hypotheses about differences between the means of two sets of data. You can:

- test the significance of a set of difference scores;
- test the significance of the difference between the means of two independent samples;
- recognize when you have independent samples and when paired data which result in difference scores.

Now try these review problems. Table I on pp. 253–256 lists any formulas you may need for reference.

1. You are studying problem-solving performance, using time as a measure. With the kind of problem you are using the distributions you obtain are typically bimodal. Your subjects either solve the set of problems quickly or take a long time, but only rarely do they take an intermediate amount of time. You select a sample of 10 subjects and give them instruction that you believe will reduce their mean time for solving the problem set. You wish to compare their performance with that of another sample of 10 who did not receive the instruction. Outline an appropriate statistical test and suggest changes in the plan of the experiment if necessary. Use a 5% significance level.

2. A laboratory owns two precision measuring devices. The director suspects that there is a slight difference in calibration between the two, so that one of them (he doesn't know which) tends to give slightly higher readings than the other. He proposes to check the two devices by taking readings of 50 objects on both machines. Thus he will have readings of objects 1–50 on Machine A and readings of objects 1–50 on Machine B. Outline on appropriate test at the 5% significance level and suggest changes in the plan if necessary.

3. From appropriately selected samples two sets of IQ scores are obtained as summarized below:

Group 1	Group 2
$n = 16$	$n = 14$
$\bar{x} = 107$	$\bar{x} = 112$
$s = 10$	$s = 8$

Is there a significant difference between the two groups? Use the 5% significance level.

Answers

To review a problem, study the frames indicated after the answer.

1. You cannot perform a statistical test as the experiment is planned. The samples are too small to use a z test, and to use a t test you must be able to assume that the populations are approximately normally distributed. Your experience indicates that you cannot make this assumption. The solution is to use samples of 30 or more. With larger samples the statistical test is the following:

Null hypothesis	$\mu_1 = \mu_2$
Alternative	$\mu_1 < \mu_2$
Significance level	$\alpha = 0.05$
Critical region	$z \leq -1.65$

The formula for z is

$$z = \frac{\bar{x}_1 - \bar{x}_2}{\sqrt{s_1^2/n_1 + s_2^2/n_2}}$$

See frames 24 to 37 and 38 to 62.

2. You must use difference scores in this case because the samples are not independent. Since you have 50 difference scores, use z.

Null hypothesis	$\mu = 0$
Alternative	$\mu \neq 0$
Significance level	$\alpha = 0.05$
Critical region	$z \leq -1.96$ or $z \geq +1.96$

The formula for z is

$$z = \frac{\bar{x} - \mu}{s/\sqrt{n}} = \frac{\bar{x} - 0}{s/\sqrt{50}} = \frac{\bar{x}}{s/\sqrt{50}}$$

See frames 1 to 23 and 38 to 62.

3. The samples are small, so you must use t. In the absence of evidence to the contrary it is reasonable to assume that the populations of IQ scores are approximately normally distributed.

Null hypothesis $\mu_1 = \mu_2$
Alternative $\mu_1 \neq \mu_2$
Significance level $\alpha = 0.05$
Critical region $t \leq -2.05$ or $t \geq +2.05$
                        (df $= 16 + 14 - 2 = 28$)

The formula for t is

$$t = \frac{\overline{x}_1 - \overline{x}_2}{\sqrt{s^2/n_1 + s^2/n_2}}$$

$$s^2 = \frac{(n_1 - 1)\,s_1^2 + (n_2 - 1)\,s_2^2}{n_1 + n_2 - 2}$$

$$s^2 = \frac{15(100) + 13(64)}{16 + 14 - 2} = \frac{1500 + 832}{28} = 83.3$$

$$t = \frac{107 - 112}{\sqrt{83.3/16 + 83.3/14}} = \frac{-5}{\sqrt{5.21 + 5.95}} = \frac{-5}{3.35} = -1.49$$

The difference between the two groups is not significant. *See frames 31 to 37 and 38 to 62.*

CHAPTER SIX

The Difference Between Two Variances or Several Means

The difference between two variances can be studied using another sampling distribution called the F distribution. The approach should, by now, be familiar. You compute a value for F using the data from your samples and use the sampling distribution table to determine if the value is in the critical region.

A related technique called *analysis of variance* allows you to consider data from several samples at the same time. You try to distinguish systematic differences between sample groups from the chance variation found within each group. Analysis of variance requires you to compute an F value.

When you have completed this chapter, you will be able to:

● test the significance of the difference between two sample variances;
● perform an analysis of variance to determine if the differences among the means of a group of samples are statistically significant;
● determine if given data meet the assumptions of these tests.

COMPARISON OF TWO VARIANCES

Just as two means can be compared by using the sampling distribution of t, two variances can be compared by using the sampling distribution of F. The procedure is very simple.

1. The variability of a population or a sample can be described by its standard deviation or by its variance. The variance is simply the square of the standard deviation; for example, if σ for a population is 6, the variance of the population

    is _____.

    _

    36

2. If s for a sample is 5, the variance of the sample is _____.

    _

    25

3. | The formula for the F ratio is
$$F = \frac{s_1{}^2}{s_2{}^2}$$

The usual way to compare the variability of two samples is to divide the larger variance by the smaller variance. The result of this computation is called an F ratio; for example, if the variance of one sample is 25 and the variance of another sample is 4, the F ratio is 25/4 or 6.25. If the variance of one sample is 36 and the variance of another is 9, the F ratio is _____.

- -

$36/9 = 4.0$

4. If one sample has a variance of 5 and the other a variance of 25, the F ratio is _____.

- -

$25/5 = 5.0$

5. An F ratio close to 1.0 indicates that the two samples have (similar/different) variances.

- -

similar

6. Two samples with very different variances would result in an F ratio that is (close to 1.0/large).

- -

large

7. In computing an F ratio, you normally divide the (larger/smaller) variance by the (larger/smaller) variance.

- -

larger
smaller

8. If two samples are drawn at random from populations with the same variance, their F ratio is most likely to be close to 1.0. A large F ratio will be relatively unlikely. For normally distributed populations it is possible to construct theoretical sampling distributions for F. This fact can be used to perform a statistical test of a hypothesis about the difference between the variance of two samples; for example, a researcher believes that stress will increase the variability of a sample's test scores. He administers the same test to two groups of 25 college applicants. One group is told that this test will be important in determining whether they will be admitted. The other group is told that the test is being administered for research purposes only. His statistical test is as follows:

Null hypothesis $\qquad$ $\sigma_1 = \sigma_2$
Alternative $\qquad$ $\sigma_1 > \sigma_2$
Significance level $\qquad$ $\alpha = 1\%$
Critical region $\qquad$ $F > 2.66$ (This is obtained from a table, as you will soon see.)

His results are

Group 1 (stress)	Group 2 (no stress)
$n = 25$	$n = 25$
$\bar{x} = 120$	$\bar{x} = 110$
$s = 20$	$s = 10$
$s^2 = 400$	$s^2 = 100$

Compute F. $F = $ _____.

- -

$F = 4.0$

9. Do the results support the researcher's theory?

- -

Yes

10. The tables used to establish a critical region for this test will tell you that a given value of F would occur by chance no more than 5% of the time, for example, or no more than 1% of the time. You will find such a table on page 271. Look at it. What is its title?

- -

"Critical Points of the F Distribution"

11. To use the F table you must know the degrees of freedom for both s_1^2 and s_2^2. As in the t test, df $= n - 1$.

As an example, $s_1^2 = 36$ $s_2^2 = 25$
$n_1 = 15$ $n_2 = 21$

The degrees of freedom for s_1^2 are 14. The degrees of freedom for s_2^2 are _____.

– –

20

12. Look at the F table and locate the area on the table that corresponds to 14 degrees of freedom for s_1^2 and 20 for s_2^2. You should find two values of F in this area. What are they?

– –

2.23 and 3.13

13. According to the table, how often would you expect an F of 3.13 or greater to occur by chance with these sample sizes?

– –

1% of the time

14. 5% of the time chance would produce an F greater than _____.

– –

2.23

15. Assume for a moment that you are operating a frozen grapefruit juice factory. To buy fruit at favorable prices you must buy the grower's entire crop, but unusually large or unusually small fruit can jam your machinery and must be eliminated before squeezing. For this reason you would like to buy crops that are relatively uniform in size; that is, you prefer crops with a (large/small) variance.

– –

small

16. Two growers offer you their crops. Grower A asks a slightly higher price than Grower B, but he says that his grapefruit are more uniform in size. To check this assertion you ask for a random sample of each crop. Each grower sends you a crate of 25 grapefruit. You measure the grapefruit in each sample and obtain the following information:

   (a)   The size of the fruit is approximately normally distributed for both samples.
   (b)   For Grower A the mean diameter of the fruit is 4.5 in., with a standard deviation of 0.5 in.
   (c)   For Grower B the mean diameter of the fruit is 4.5 in., with a standard deviation of 1 in.

   State an appropriate null hypothesis and alternative for a statistical test.

   $-------------------------$

   Null hypothesis $\sigma_1 = \sigma_2$
   Alternative $\sigma_1 < \sigma_2$

17. For a significance level of 5% what is the critical region? (Use the table.)

   $F \geq$ _____.

   $-------------------------$

   $F \geq 1.98$

18. Compute F. Are the results significant?

   $-------------------------$

   $$F = \frac{1^2}{0.5^2} = \frac{1}{0.25} = 4.0$$

   The results are significant. The fruit from Grower A is more uniform in size.

19. In the cases you have considered so far the alternative called for a one-tailed statistical test; that is, your theory specified which of the two variances would be larger. Sometimes your alternative is two-tailed. It says only that $\sigma_1 \neq \sigma_2$ without stating which one will be larger. In this case you will use whichever variance happens to be larger as the numerator of the F ratio but you must double the probabilities when you use the F table. The 5% values become 10% and the 1% become ___% values.

- -

2%

20. You have been given the following statistics on two samples. From experience you are quite certain that the populations are normally distributed but you suspect that the two populations are not equally variable.

Sample 1	Sample 2
$n = 10$	$n = 10$
$x = 8.8$	$x = 12.7$
$s = 12$	$s = 5$

What would be an appropriate null hypothesis and alternative for investigating this matter?

- -

Null hypothesis $\sigma_1 = \sigma_2$
Alternative $\sigma_1 \neq \sigma_2$

21. Using a 2% significance level, establish a critical region.

- -

Critical region $F \geq 5.35$

22. Compute F.

- -

$$F = \frac{12^2}{5^2} = \frac{144}{25} = 5.76$$

23.   Will a t test be appropriate in this case?

_ _

No. The difference between the variance of the two samples is exceptionally large. It is not likely that they came from populations with the same variance.

ANALYSIS OF VARIANCE

The F ratio can be used to test the null hypothesis that a number of samples all come from populations with the same mean. The procedure used for this test is called analysis of variance.

24.   In the analysis of variance we use the difference among sample means to estimate the variance of the population. We also make a separate estimate of the population variance based only on the differences among individuals within each sample. If the samples all come from populations with the same mean, the differences between sample means will be relatively (large/small).

_ _

small

25.   If the differences between sample means are relatively large, we will conclude that the mean of the populations (is/is not) the same.

_ _

is not

26.   The estimate of variance based on the differences between the means of groups is called "between groups variance." The estimate of variance based only on the differences between individuals is called "within groups variance" or "error variance." To reject the null hypothesis between groups variance must be large

compared with _____.

_ _

within groups variance or error variance

27. Let us see how between groups variance and within groups variance can be estimated. The purpose of this explanation is to show you the basic logic of the test. You will learn simplified procedures for computation later. A researcher believes that the color of a toy will affect how long children will play with it. From a population of preschool children he obtains four sample groups of 10 children each. Using the same stuffed animal, but in a different color for each sample group, he observes how many minutes each child in each sample group spends playing with the toy during a 10-minute session. His null hypothesis is that $\mu_1 = \mu_2 = \mu_3 = \mu_4$. The alternative is that not all the means are equal; that is, that the color of the toy does make a difference. His data look like this:

Group 1 (red giraffe)		Group 2 (yellow giraffe)		Group 3 (green giraffe)		Group 4 (blue giraffe)	
1		3		2		5	
2	$n_1 = 10$	2	$n_2 = 10$	4	$n_3 = 10$	3	$n_4 = 10$
5		6		2		1	
7	$\bar{x}_1 = 3.4$	3	$\bar{x}_2 = 5.0$	1	$\bar{x}_3 = 2.4$	2	$\bar{x}_4 = 2.5$
6		2		2		1	
1	$s_1^2 = 4.5$	8	$s_2^2 = 5.6$	3	$s_3^2 = 1.2$	3	$s_4^2 = 1.8$
2		7		4		4	
2		5		1		2	
4		6		3		3	
4		8		2		1	

All groups combined

$N = 40$
$\bar{x}_T = 3.3$
$s_T^2 = 4.1$

The between groups variance estimate is based on the differences between (individuals/group means).

_ _

group means

28. Consider the distribution of the four group means. This distribution is

(a) a population distribution
(b) a sample distribution
(c) a sampling distribution

_ _

(c) sampling distribution

29. As you recall from our discussion of the central limit theorem,

$$\sigma_{\overline{x}} = \frac{\sigma}{\sqrt{n}}$$

that is, the standard deviation of the sampling distribution of x depends on two things, σ and n.

What does σ stand for?
What does n stand for?

_ _

σ is the standard deviation of the population.
n is the sample size.

30. At the moment we are interested in computing s^2, an estimate of the variance of the population. We can find it by using a formula based on the formula above. Using s in the place of σ to indicate that this is an estimate, we can say $s_{\overline{x}} = s/\sqrt{n}$. Then $s_{\overline{x}}^2 = s^2/n$, and $s^2 = ns_{\overline{x}}^2$. What is the size of the samples that make up our sampling distribution? $n =$ _____

_ _

10

31. $s_{\overline{x}}^2$ is simply the variance of the following numbers:

3.4, 5.0, 2.4, and 2.5.

What are these numbers?

_ _

The means of the four samples.

32. The variance of the sample means works out to be 1.45. (If you would like practice, check it for yourself.)

On the basis of this information, what is the variance of the total population? $s^2 = ns_{\bar{x}}^2 = $ _____.

— —

$10(1.45) = 14.5$

33. 14.5 is the (between/within) groups variance estimate.

— —

between

34. The between groups variance estimate is based on the variance of (individual scores/group means).

— —

group means

35. The within groups variance estimate is based on the variances of the individual scores within sample groups. What are the variances of individual scores within the sample groups?

$s_1^2 =$

$s_2^2 =$

$s_3^2 =$

$s_4^2 =$

- -

$s_1^2 = 4.5$

$s_2^2 = 5.6$

$s_3^2 = 1.2$

$s_4^2 = 1.8$

36. Are these variances affected by the differences between groups?

- -

No

37. The variance of each of these groups is an estimate of the total population variance. To obtain an even better estimate we can average all four group variances. This is our within groups variance estimate. Compute the within groups variance estimate for this example.

- -

$$\frac{4.5 + 5.6 + 1.2 + 1.8}{4} = 3.28$$

38. Compute the F ratio.

$$F = \frac{\text{between groups variance}}{\text{within groups variance}} =$$

- -

$$F = \frac{14.5}{3.28} = 4.42$$

39. To evaluate this F ratio we still need to know the degrees of freedom for the between groups and within groups variance. The between groups variance estimate is based on four means. Degrees of freedom for this variance are $4 - 1 = 3$. The within groups variance is an average of several variances and the degrees of freedom for it are the total of the degrees of freedom for each of the groups. What are the degrees of freedom for one group of 10?

- -

9

40. What are the total degrees of freedom for four groups of 10?

$4 \times 9 = 36$

41. The degrees of freedom for s_b^2 (between groups variance) are _____. The degrees of freedom for s_w^2 (within groups variance) are _____. For a 1% significance level the critical region is $F \geq$ _____.

3
36
$F \geq 4.38$

42. Will the experimenter reject the null hypothesis?

Yes, because the F ratio is in the critical region. He can conclude that the color of the toy does make a difference in its attractiveness.

43. In the example you have just considered the sample groups were of equal size. When the sample groups are not of equal size, it is necessary to weight the means and variances of the sample groups according to their sizes. The computations are more easily done when you use "sums of squares" instead of means and variances. A sum of squares is the numerator (upper term) of a variance estimate. Circle the part of the following formula that represents the sum of squares.

$$s^2 = \frac{\Sigma(x - \overline{x})^2}{n - 1}$$

$$s^2 = \frac{\boxed{\Sigma(x - \overline{x})^2}}{n - 1} \longleftarrow \text{sum of squares}$$

44. The between groups sum of squares is the sum of squares you will use to compute the between groups variance estimate. The within groups sum of squares is the sum of squares you will use to compute the within groups variance estimate. If you consider all the groups combined as a single large sample, you can also compute a total sum of squares. To compute the variance of all the observations combined as a single sample you would use the _____ sum of squares.

- -

total

45. Adding the between groups sum of squares to the within groups sum of squares will always give you the total sum of squares. This fact can simplify computations; for example, if you know that the total sum of squares is 75 and the between groups sum of squares is 25, the within groups sum of squares must be ___.

- -

50

In practical work the mathematical procedures for analysis of variance are normally done by computer. To understand the terminology and to appreciate what you are missing you nevertheless ought to work through a few problems by hand. If you have no tolerance for computation, at least read through the explanations of the following problems step by step. You will need this background to understand computer output.

46. To perform an analysis of variance of data from varying size groups start by setting up a table like this:

Group 1		Group 2		Group 3		Group 4		Total
x	x^2	x	x^2	x	x^2	x	x^2	
3	9	2	4	3	9	5	25	
3	9	4	16	5	25	6	36	$\Sigma x_T = 125$
5	25	4	16	5	25	6	36	$\Sigma x_T^2 = 799$
6	36	6	36	6	36	7	49	$N = 22$
8	64	9	81	7	49			
		10	100	7	49			
				8	64			
25	143	35	253	41	257	24	146	
$n_1 = 5$		$n_2 = 6$		$n_3 = 7$		$n_4 = 4$		

From this table you can read all the information you will need to complete the analysis of variance. Notice that there are two columns for each group. The first column lists the scores for the group. The second column lists

_____.

- -

the scores squared

47. Do you see any columns for $(x - \overline{x})$ or $(x - \overline{x})^2$?

- -

No

48. What symbol is used to indicate the number of observations in group 1?

- -

n_1

49. What symbol is used to indicate the total number of observations in all groups combined?

- -

N

50. What symbol is used to indicate the grand total of all the observations?

_ _

Σx_T

51. You can compute all the necessary sums of squares, using only Σx and Σx^2 for the various groups. The formulas are as follows:

Total sum of squares

$$\Sigma x_T^2 - \frac{(\Sigma x_T)^2}{N}$$

Degrees of freedom for the total sum of squares depend on the total size of all samples combined: df $= N - 1$.

Between groups sum of squares

$$\frac{(\Sigma x_1)^2}{n_1} + \frac{(\Sigma x_2)^2}{n_2} + \cdots - \frac{(\Sigma x_T)^2}{N}$$

Degrees of freedom for the between groups sum of squares depend on the number of groups. We use g to indicate the number of groups.

df $= g - 1$

Within groups sum of squares

total sum of squares $-$ between groups sum of squares or

$$\Sigma x_1^2 - \frac{(\Sigma x_1)^2}{n_1} + \Sigma x_2^2 - \frac{(\Sigma x_2)^2}{n^2} + \cdots$$

Degrees of freedom for the within groups sum of squares depend on the degrees of freedom for each of the individual groups. df $= (n_1 - 1) + (n_2 - 1) + \ldots$. A simpler way to calculate this is df $= N - g$.

In the example above what value from the table corresponds to Σx_T?

_ _

125

52. What value corresponds to Σx_T^2?

_ _

799

53. What value will correspond to $(\Sigma x_T)^2$?

15625, 125^2

54. What value corresponds to Σx_3^2?

257

55. What value corresponds to $(\Sigma x_1)^2$?

625, 25^2

56. Calculate the total sum of squares for the data in frame 46. Refer back to the formulas if you wish.

$$799 - \frac{(125)^2}{22} = 799 - 710.2 = 88.8$$

57. Calculate the between groups sum of squares.

_ _

$$\frac{(25)^2}{5} + \frac{(35)^2}{6} + \frac{(41)^2}{7} + \frac{(24)^2}{4} - \frac{(125)^2}{22} = 125 + 204.2 + 240.1 + 144 - 710.2 = 3.1.$$

58. Calculate the within groups sum of squares.

_ _

$88.8 - 3.1 = 85.7$

59. Complete the following table:

	Sum of Squares	df	Variance Estimate
Total			...
Between groups			
Within groups			

_ _

	Sum of Squares	df	Variance Estimate
Total	88.8	21	...
Between groups	3.1	3	1.03
Within groups	85.7	18	4.76

60. $F = \dfrac{s_b^2}{s_w^2}$

What is the F ratio for these data? $F = $ _____.

- -

$\dfrac{1.03}{4.76} = 0.22$

61. Let's review the meaning of between groups and within groups variance. The between groups variance estimate is based on the means of the groups. The within groups variance estimate is based on the variances of the individual groups. Which variance estimate reflects *only* the chance variations involved in drawing a sample?

- -

Within groups. The between groups variance also reflects the intentional difference between groups.

62. Which variance estimate reflects the intentional differences between groups?

- -

Between groups

63. If the between groups variance is less than the within groups variance, you can say that the differences between groups are (large/small) compared with the chance variation involved in drawing a sample.

— —

small

64. What does the within groups variance estimate reflect?

— —

Chance, the chance variations involved in drawing a sample.

65. What does the between groups variance estimate reflect?

— —

The differences between groups

66. If s_b^2 is smaller than s_w^2, can the results be significant?

— —

No

67. The following scores represent the performance on a trouble-shooting test of small random samples of people with different educational backgrounds. Does educational background appear to make a difference in the performance on this test? Use a significance level of 1%.

	Scores
Group 1 (high school)	1, 3, 4
Group 2 (technical institute)	4, 5, 6, 6, 7, 8, 9
Group 3 (college)	2, 3, 3, 4

Suggestions:

(a) Set up a table for the data like the one in frame 46.

(b) Compute the sums of squares and the variance estimates and summarize them in a table like the one in frame 59.

(c) Calculate F and refer to the table to determine if it is significant.

– –

Group 1		Group 2		Group 3		Total
x	x^2	x	x^2	x	x^2	
1	1	4	16	2	4	
3	9	5	25	3	9	$\Sigma x_T = 65$
4	16	6	36	3	9	$\Sigma x_T^2 = 371$
		6	36	4	16	$N = 14$
		7	49			
		8	64			
		9	81			
8	26	45	307	12	38	
$n_1 = 3$		$n_2 = 7$		$n_3 = 4$		

	Sum of Squares	df	Variance Estimate
Total	69.2	13	...
Between groups	44.8	2	22.40
Within groups	24.4	11	2.22

$$F = \frac{22.40}{2.22} = 10.1$$

Critical region $F \geq 7.20$

The null hypothesis is rejected. Educational background does make a difference.

68.   Look at this printout of an analysis of variance calculated by computer.

SOURCE	DF	SUM SQUARES	F RATIO	PROB
BETWEEN GROUPS	5	361.3217	11.526	0.000
WITHIN GROUPS	494	3097.3463		
TOTAL	499	3458.6680		

Note the degrees of freedom; how many groups were used in the experiment? _____ How many individuals were studied? _____ Are the differences among groups significant at the 1% level? (Yes/No)

- -

6 groups (df = 6 − 1)
500 individuals (df = 500 − 6)
Yes, the results are significant.

WHEN TO USE ANALYSIS OF VARIANCE

Certain assumptions apply to the analysis of variance. Only when these assumptions are met is it appropriate to use analysis of variance to test your hypothesis.

69. Analysis of variance and the t test for the difference between two means are based on similar mathematical derivations. They depend on the same assumptions and they yield the same results when applied to two samples. Both tests require what two assumptions?

_ _

The population distributions are normal.
The population variances are equal.

70. If you know that the distribution of family incomes is strongly skewed, will it be appropriate to use analysis of variance to study the effect of level of education on family income?

_ _

No. You cannot assume that the populations are normally distributed.

71. Will it be appropriate to use a t test?

_ _

No. The t test also assumes normally distributed populations.

72. If comparisons are to be made among several different samples, t tests are not appropriate; for example, if you are comparing four sample means, six different two-way comparisons are possible. If you select a significance level of 5%, each individual comparison has a 5% chance of a Type I error. For all six comparisons you have substantially more than a 5% chance of at least one Type I error. The analysis of variance avoids this problem because it makes only one comparison (between s_b^2 and s_w^2). Therefore, when you are making comparisons among a number of samples, the appropriate test to use is (a t test/ analysis of variance).

_ _

analysis of variance

73. The following is a brief summary of the adjustment scores of mental patients treated with three different drugs. Is analysis of variance an appropriate statistical technique to apply in this case?

Group 1	Group 2	Group 3
$\bar{x} = 5$	$\bar{x} = 10$	$\bar{x} = 4$
$s = 4$	$s = 23$	$s = 5$
$n = 10$	$n = 33$	$n = 45$

– –

No

74. What assumption is violated by the above data?

– –

The variances are very unequal ($s_1^2 = 16$, $s_2^2 = 529$, $s_3^2 = 25$). In this case, you might be able to restructure your data and apply the *Chi square* test discussed in Chapter 8.

REVIEW PROBLEMS

If you have successfully completed this chapter you can now use the F distribution to perform statistical tests. You can:

- test the significance of the difference between two sample variances;
- perform an analysis of variance to determine if the differences among a number of sample means are statistically significant;
- recognize situations where F tests should not be used because the data does not meet the assumptions of the mathematical model.

Now try these review problems. Table I on pp. 253–256 lists any formulas you may need for reference.

1. A number of students are assigned randomly to three classes with three different teaching methods. The following statistics summarize the performance of the three groups on a comprehensive final exam. Can you perform an analysis of variance with these data? What assumptions are involved?

Group 1	Group 2	Group 3
$n = 10$	$n = 11$	$n = 8$
$\bar{x} = 89$	$\bar{x} = 75$	$\bar{x} = 90$
$s^2 = 100$	$s^2 = 81$	$s^2 = 64$
$s = 10$	$s = 9$	$s = 8$

2. The situation is the same as in problem 1 but the data are different. Can you perform an analysis of variance? What assumptions are involved?

Group 1	Group 2	Group 3
$n = 10$	$n = 11$	$n = 8$
$\overline{x} = 90$	$\overline{x} = 86$	$\overline{x} = 70$
$s^2 = 144$	$s^2 = 25$	$s^2 = 16$
$s = 12$	$s = 5$	$s = 4$

3. You believe that a new production procedure will reduce the variability in size of molded plastic parts. You obtain samples of parts molded under the new and old procedures and measure them as summarized below. Outline an appropriate statistical test at the 5% significance level. Is the difference between the variability of the two samples significant? Is the difference between their means significant?

Group 1	Group 2
$n = 15$	$n = 25$
$\overline{x} = 20.00$	$\overline{x} = 21.00$
$s^2 = 0.0625$	$s^2 = 1.00$
$s = 0.25$	$s = 1.00$

4. Perform an analysis of variance of the following data. Use a 5% significance level.

Group 1	Group 2	Group 3
1	2	3
2	3	4
2	4	4
2	4	5
3	5	
	6	

Answers

To review a problem, study the frames indicated after the answer.

1.   Yes, an analysis of variance is possible. You must be prepared to assume that the populations from which you are sampling are normally distributed. To interpret the results you must also be prepared to assume that the teaching method is the only reasonable explanation of the differences between groups. *See frames 69 to 74 and frames 26 to 31 in Chapter 4.*

2.   No, you cannot perform an analysis of variance. One of the assumptions for analysis of variance is that the populations sampled have equal variances. The variances of these three samples are too different to allow this assumption. The variance of group 1 is nine times as great as that of group 3 and almost six times as great as that of group 2. *See frames 1 to 23 and 69 to 74.*

3.   Null hypothesis $\quad\quad \sigma_2 = \sigma_1$
     Alternative $\quad\quad\quad \sigma_2 > \sigma_1$
     Significance level $\quad \alpha = 0.05$
     Critical region $\quad\quad F \geq 2.35 \,(\mathrm{df} = 24, 14)$

$$F = \frac{s_2^2}{s_1^2} = \frac{1.00}{0.0625} = 16.0$$

The difference in variances is significant. With this substantial difference in variances, the accuracy of a t test would be in doubt. *See frames 1–23.*

4.

x_1	x_1^2	x_2	x_2^2	x_3	x_3^2	
1	1	2	4	3	9	
2	4	3	9	4	16	$\Sigma x_T = 50$
2	4	4	16	4	16	$\Sigma x_T^2 = 194$
2	4	4	16	5	25	$N = 15$
3	9	5	25			
		6	36			
$\overline{10}$	$\overline{22}$	$\overline{24}$	$\overline{106}$	$\overline{16}$	$\overline{66}$	

$$n_1 = 5 \qquad n_2 = 6 \qquad n_3 = 4$$

Total sum of squares

$$\Sigma x_T^2 - \frac{(\Sigma x_T)^2}{N} = 194 - \frac{(50)^2}{15} = 194 - \frac{2500}{15} = 194 - 166.7 = 27.3$$

$df = N - 1 = 14$

Between groups sum of squares

$$\frac{(\Sigma x_1)^2}{n_1} + \frac{(\Sigma x_2)^2}{n_2} + \frac{(\Sigma x_3)^2}{n_3} - \frac{(\Sigma x_T)^2}{N} = \frac{(10)^2}{5} + \frac{(24)^2}{6} + \frac{(16)^2}{4} - 166.7 = 20 +$$
$$+ 96 + 64 - 166.7 = 13.3$$

$df = g - 1 = 2$

Within groups sum of squares

total $-$ between $= 27.3 - 13.3 = 14.0$

$df = N - g = 12$

	Sum of Squares	df	Variance Estimate	F
Total	27.3	14		
Between groups	13.3	2	6.65	5.68
Within groups	14.0	12	1.17	

The critical region is $F \geq 3.89$ for a 5% significance level. The difference between groups is significant. *See frames 24 to 67.*

CHAPTER SEVEN
The Relation Between Two Sets of Measures

So far, we have been concerned with only one kind of data at a time. In this chapter you will learn some ways of studying the relationships between two different measures. For example, you may wonder if the height of a tree is related to its age, or if wealth and happiness go together. A *scattergram* can be used to display graphically the relation between two different measures in a sample. It is also possible to summarize the relation between two measures quantitatively, using a *correlation coefficient*. And, given certain assumptions, it is possible to test the significance of a correlation coefficient by referring to a sampling distribution.

If two measures are related, it is possible to use one measure for an individual to predict the other. This is done by means of a *regression equation*. For example, if wealth and happiness are related, you might be able to apply a regression equation to a person's happiness rating to determine his or her approximate wealth.

When you have completed this chapter, you will be able to:

- construct and interpret a scattergram;
- compute and interpret the correlation coefficient r;
- test the null hypothesis that two measures have zero correlations;
- use a regression equation to predict the value of one measure on the basis of a related measure.

SCATTERGRAMS

Often we are interested in the relationship between two different measurements or observations in a population; for example, do high scores on a scholastic aptitude test tend to go with high grade point averages or are the two measures unrelated? Do rich families tend to have many children or few children, or is there no relation at all between wealth and the number of children in a family? One simple way to examine such a relationship is to draw a scattergram.

1. Look at the scattergram below. Each *x* on the scattergram represents a family in our sample. The height of the *x* on the graph represents the number of children in the family. The left-right position of the *x* represents the family income; for example, A represents a family with three children and an income of $8000.

B represents a family with _____ children and an income of _____.

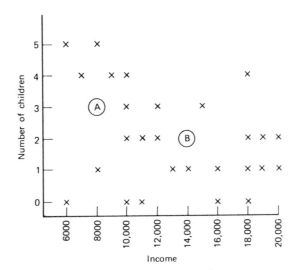

- -

two
$14,000

2. In the scattergram below circle the *x* that represents a family with no children and an income of $6000.

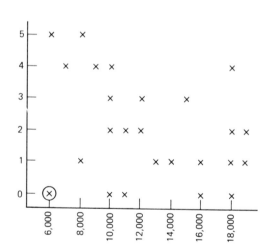

3. Looking at the scattergram, you can say that in this sample families with high incomes tend to have relatively (larger/smaller) numbers of children.

smaller

4. Construct a scattergram for the following data:

Student No.	Midterm Exam Grade	Final Exam Grade
1	65	74
2	68	73
3	71	71
4	75	80
5	75	83
6	85	85
7	88	90
8	95	94
9	97	99
10	100	98

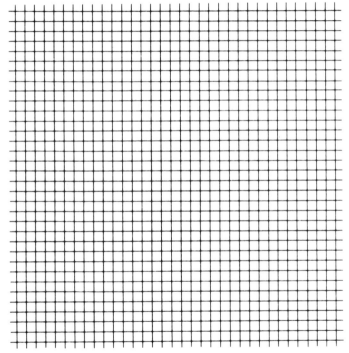

Your scattergram should look like the one below.

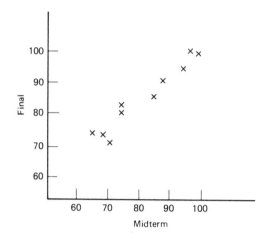

You might have reversed the axes so that the vertical dimension represented the midterm grade and the horizontal dimension, the final grade.

When one measure may be used to predict another, it is customary to represent the predictor on the horizontal dimension (the x axis).

5. In this case high scores on the midterm tend to go with (high/low) scores on the final exam.

– –

high

6. In each of the two scattergrams you have seen you could make a reasonably good representation of the relationship between the two measurements by drawing a straight line. This kind of relationship is called a linear relationship, as shown below.

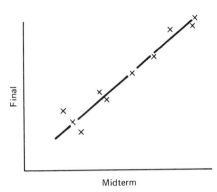

Which of the scattergrams below shows a clear linear relationship?

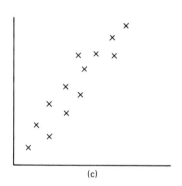

_ _

(c)

7. Sometimes a relationship between two measurements is curvilinear; that is, it is best described by a curved line. Which of the above scattergrams represents a curvilinear relationship?

_ _

(b)

8. A relationship that can be described by a straight line is called a _____ relationship.

_ _

linear

9. A relationship that is best described by a curved line is called a _____
relationship.

_ _

curvilinear

10. When there is no relationship between two variables, the scattergram will look
something like this.

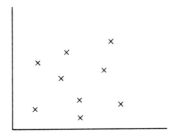

The stronger the relationship, the more closely the points on your scattergram
will approach some linear or curvilinear pattern. Which of the scattergrams
below represents the stronger relationship between two variables?

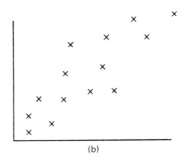

(a)   (b)

_ _

(a)

11. Which of these two scattergrams represents the stronger relationship between two variables?

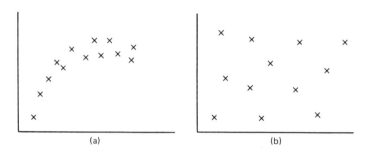

(a)

- -

(a)

CORRELATION COEFFICIENTS

If two measures have a linear relationship, it is possible to describe how strong the relationship is by means of a statistic called a correlation coefficient. The symbol for the correlation coefficient is *r*. The symbol for the corresponding population parameter is ρ (the Greek letter "rho").

There are various forms of correlation coefficient for specialized types of data. We explain only the basic type, which is called the Pearson product-moment correlation.

12. Match the explanations of correlation coefficients to the appropriate illustrations below.

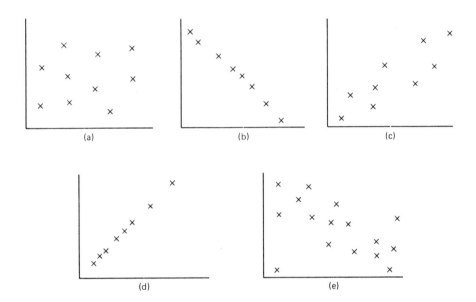

Correlation Coefficient		Figure
$r = +1.0$	All scores are exactly on the line. High scores on one measure go with high scores on the other.	_____
r between +1.0 and 0 (e.g., +0.5)	High scores on one measure tend to go with high scores on the other but the relationship is not perfect.	_____
$r = 0$	No relationship at all between the two measures.	_____
r between 0 and −1.0 (e.g., −0.3)	High scores on one measure tend to go with low scores on the other but the relationship is not perfect.	_____
$r = -1.0$	All scores are exactly on the line. High scores on one measure go with low scores on the other.	_____

– –

(d)
(c)
(a)
(e)
(b)

13.

> The basic formulas for the correlation coefficient are
>
> $$\rho = \frac{1}{N} \Sigma \frac{(x - \mu_x)}{\sigma_x} \cdot \frac{(y - \mu_y)}{\sigma_y}$$
>
> $$r = \frac{1}{n-1} \Sigma \frac{(x - \bar{x})}{s_x} \cdot \frac{(y - \bar{y})}{s_y}$$

Let us apply this to a simple example. Suppose you are interested in the relation between the age of garden snakes (in months) and their length (in inches). You obtain a sample of three garden snakes of known age and you measure them with the following results:

Snake No.	Age (x)	Length (y)
1	1	4
2	2	6
3	3	8

$\bar{x} = 2$ $\bar{y} = 6$
$s_x = 1$ $s_y = 2$

Since this is a sample, the formula we must use for a correlation coefficient

is $r = $ _____.

- -

$$r = \frac{1}{n-1} \Sigma \frac{(x - \bar{x})}{s_x} \cdot \frac{(y - \bar{y})}{s_y}$$

14. For each snake we must compute $(x - \bar{x})/s_x \cdot (y - \bar{y})/s_y$. Note carefully that $(x - \bar{x})/s_x$ tells us how many standard deviations a given snake's age is away from the mean. It is a form of z score. What does $(y - \bar{y})/s_y$ tell us?

- -

How many standard deviations the snake's length is away from the mean.

15. Complete this table to find $(x - \bar{x})/s_x \cdot (y - \bar{y})/s_y$ for each snake.

Age x	$(x - \bar{x})$	$\dfrac{x - \bar{x}}{s_x}$	Length y	$(y - \bar{y})$	$\dfrac{y - \bar{y}}{s_y}$	$\dfrac{(x - \bar{x})}{s_x} \cdot \dfrac{(y - \bar{y})}{s_{\bar{y}}}$
1			4			
2			6			
3			8			

$\bar{x} = 2$ $\bar{y} = 6$

$s_x = 1$ $s_y = 2$

- -

Age x	$(x - \bar{x})$	$\dfrac{x - \bar{x}}{s_x}$	Length y	$(y - \bar{y})$	$\dfrac{(y - \bar{y})}{s_y}$	$\dfrac{(x - \bar{x})}{s_x} \cdot \dfrac{(y - \bar{y})}{s_y}$
1	-1	-1	4	-2	-1	$+1$
2	0	0	6	0	0	0
3	$+1$	$+1$	8	$+2$	$+1$	$+1$

16. For this sample what is $\Sigma(x - \bar{x})/s_x \cdot (y - \bar{y})/s_y$? Add up the sum in your table.

- -

$+1 + 0 + 1 = +2$

17. Compute r.

$$r = \frac{1}{n-1}\Sigma\frac{(x - \bar{x})}{s_x} \cdot \frac{(y - \bar{y})}{s_y} = \underline{\hspace{2cm}}$$

- -

$$r = \frac{1}{3-1} \cdot 2 = +1.0$$

18. A correlation coefficient of $+1.0$ means that for every pair of observations $(x - \bar{x})/s_x$ is exactly equal to $(y - \bar{y})/s_y$. When a snake's length is two standard deviations above the mean, its age is _____.

- -

Two standard deviations above the mean

19. If the correlation coefficient is $+1.0$, information about one measurement tells you precisely what the other measurement must be; for example, suppose that instead of three garden snakes, you have a large population. For age the mean is 12 months, with a standard deviation of 4 months. For length the mean is 25 in., with a standard deviation of 8 in. The correlation between age and length is $\rho = +1.0$. You are going to pick one snake at random from this population. Can you predict accurately what its age will be?

- -

No. Your best guess would be 12 months, but the odds are small that you will be precisely correct.

20. Can you predict accurately what its length will be?

    _____

    No

21. You select a snake and determine its age. Can you predict accurately what its length will be?

    _____

    Yes. If you know how many standard deviations above or below the mean its age is, you know that its length must be the same number of standard deviations above or below the mean.

22. If a correlation coefficient is -1.0, you can also predict one measure precisely by knowing the other. When $r = -1.0$, $(x - \overline{x})/s_x = -(y - \overline{y})/s_y$. An x score two standard deviations above the mean must be paired with a y score two standard deviations (above/below) the mean.

    _____

    below

23. You are studying reaction time in middle-aged drivers. The correlation between age and reaction time is -0.96. If you know the age of a driver from this population, can you state precisely what his reaction time will be?

    _____

    No. The relationship is strong but not perfect.

24. On the basis of this correlation coefficient, describe in general terms how age is related to reaction time in the population.

    _____

    Higher ages tend to go with longer reaction times.

25. If you know a driver's age, can you make a more accurate prediction of his reaction time than if his age were unknown?

_ _

Yes

26. In the case of the garden snakes we can say that the age of the snakes explains all of the variance in their lengths; that is, knowing the ages, we can predict their lengths precisely. In the case of the drivers does age account for all of the variance in their reaction times?

_ _

No

27. The square of the correlation coefficient tells us exactly what percentage of the variance of y is explained by x; for example, the percentage of the variance in reaction times explained by differences in age in our group of drivers is r^2 or _____%.

_ _

92%

28. If the correlation between aptitude-test scores and grade-point averages in a population of students is 0.70, the aptitude-test scores explain _____% of the variance in academic performance.

_ _

49%

29. When we say that 49% of the variance in academic performance is "explained" by aptitude-test scores, it is important to remember that we are not necessarily describing a cause and effect relationship; we are only describing how one variable can be used to predict another; for example, could you use academic performance (grade-point averages) to predict aptitude-test scores?

_ _

Yes

30. The correlation between grade-point averages and aptitude-test scores is $+0.70$. Academic performance therefore "explains" $r^2 = 49\%$ of the variance in _____.

- -

aptitude-test scores

31. A researcher is able to determine the total lifetime earnings and the age at which they died for an appropriate sample of working men. The correlation between total lifetime earnings and life-span is $+0.80$. The researcher concludes that poverty is a cause of premature death. Does his statistical information support this conclusion? Why?

- -

No. The correlation coefficient does not tell which measurement is cause and which effect, or if some other factor underlies both of the measured items. Early death probably reduces total lifetime earnings. Health problems might cause both early death and low earnings. All the researcher can conclude from the correlation is that early death and low lifetime earnings go together.

COMPUTING r

For most practical purposes, correlation coefficients are calculated by computers. Nevertheless, at least once in your life you ought to compute one by hand so that you can check the honesty of your computer.

32. It is not necessary to compute $(x - \bar{x})/s_x$, $(y - \bar{y})/s_y$ for every pair of scores to compute r. In practice, less computation is required if you use the following formula:

$$r = \frac{n\Sigma(xy) - (\Sigma x)(\Sigma y)}{\sqrt{[n\Sigma x^2 - (\Sigma x)^2][n\Sigma y^2 - (\Sigma y)^2]}}$$

Courage! The formula is not so difficult as it looks. You can set up a table similar to those used for computing standard deviations or for analysis of variance. Use the information from the following table to complete the computation of r. (If you have a calculator, this is a good place to use it.)

x	x^2	y	y^2	xy
1	1	2	4	2
2	4	4	16	8
4	16	5	25	20
5	25	7	49	35
5	25	8	64	40
17	71	26	158	105

- -

$$r = \frac{5(105) - (17)(26)}{\sqrt{[5(71) - (17)^2][5(158) - (26)^2]}} = +0.97$$

33. Set up a table and compute *r* for the following data:

x	y
5	4
6	3
1	2
4	6
2	3

- -

x	x^2	y	y^2	xy
5	25	4	16	20
6	36	3	9	18
1	1	2	4	2
4	16	6	36	24
2	4	3	9	6
18	82	18	74	70

$$r = \frac{5(70) - (18)(18)}{\sqrt{[5(82) - (18)^2][5(74) - (18)^2]}} = \frac{26}{\sqrt{[86][46]}} = +0.41$$

HYPOTHESIS TESTING

As with other statistics, it is possible to draw conclusions about the correlations within a population on the basis of a sample by referring to a sampling distribution. That is, you can test the significance level of a correlation coefficient.

34. If you look at pairs of observations in a small sample, you are quite likely to find correlations by chance; for example, in the space below write the last seven digits of your social security number in column a and write your telephone number in column b. Now compute the correlation between the two sets of observations. Refer to the formulas on page 253 if you wish or use

$$r = \frac{n\Sigma(xy) - (\Sigma x)(\Sigma y)}{\sqrt{[n\Sigma x^2 - (\Sigma x)^2][n\Sigma y^2 - (\Sigma y)^2]}}$$

a (Social Security No.)		b (Telephone No.)		
x	x^2	y	y^2	xy

$\Sigma x =$ $\Sigma x^2 =$ $\Sigma y =$ $\Sigma y^2 =$ $\Sigma xy =$

- -

You almost certainly obtained some value of r other than 0. The probability that it was somewhere between $+0.75$ and -0.75 is 95%.

35. As you no doubt suspect, we can construct theoretical sampling distributions for r under certain assumptions. Specifically, if we assume that we are drawing random samples from a population in which x and y are both normally distributed and have zero correlation, we can compute the probability of obtaining various values of r. This information is presented in the table on page 275. What is the title of the table?

- -

Critical values of r.

36. We can use this table to test the null hypothesis that the population correlation ρ is zero; for example, a researcher believes that there is a correlation between the blood pressure and the amount of a certain chemical in the blood. She knows that both measures are normally distributed. From a random sample of 27 individuals she computes a correlation coefficient. What is the null hypothesis?

Null hypothesis $\rho =$ _____ .

- -

0

37. What is the alternative?

- -

$\rho \neq 0$

38. In establishing a critical region, she will consider (both ends/one end) of the sampling distribution.

- -

both ends

39. Let us use a 5% significance level. Since both ends of the sampling distribution are being considered, we must find the critical value of r that corresponds to 0.025 for our sample size. What is the critical value?

- -

0.396

40. What is the critical region?

- -

$r \geq + 0.396$ or $r \leq - 0.396$

41. If the researcher obtains a value from his sample of $r = -0.29$, can she reject the null hypothesis?

– –

No

42. Work through one more example. You believe that the speed in words per minute and error rate in errors per page of typists are negatively correlated; that is, fast typists tend to make (more/fewer) errors than slow ones.

– –

fewer

43. You will test this theory with a sample of 20 typists. Outline an appropriate statistical test of the theory, using a significance level of 1%.

Null hypothesis	_____
Alternative	_____
Significance level	_____
Critical region	_____

– –

Null hypothesis	$\rho = 0$
Alternative	$\rho < 0$
Significance level	$\alpha = .01$
Critical region	$r \leq -.516$

44. Using the table of critical values of r involves two assumptions.

(a) The two measures are distributed _____.

(b) The sample is chosen at _____.

– –

normally
random

45. To test your theory about the relation between typing speed and error rate you choose five very fast typists, five moderately fast typists, five moderately slow typists, and five very slow typists for your sample. Does this procedure meet the assumption of your statistical test? Why?

No. This is not a random sample because not all typists have an equal chance of being selected. You cannot use the table of critical values of r, but you can use other statistical procedures as you will see in a moment.

46.   It is not uncommon for researchers doing exploratory studies to collect data on a large number of variables and use a computer program to calculate all possible correlation coefficients in a search for clues to meaningful relationships. Look at the following computer printout.

PEARSON CORRELATION COEFFICIENTS

VARIABLE PAIR			VARIABLE PAIR		
FOODCOST		.2598	FOODCOST		.5798
WITH	N	45	WITH	N	45
RENTCOST	P	.042	TRNSCOST	P	.0000

VARIABLE PAIR			VARIABLE PAIR		
FOODCOST		.5469	RENTCOST		−.0111
WITH	N	44	WITH	N	45
PAYLVL	P	.000	TRNSCOST	P	.471

VARIABLE PAIR			VARIABLE PAIR		
RENTCOST		−.0249	TRNSCOST		.6858
WITH	N	44	WITH		44
PAYLVL	P	.436	PAYLVL		.049

Which correlations are significant at the 1% level? _____

FOODCOST WITH TRNSCOST and
FOODCOST WITH PAYLVL

47.   Suppose you collect data from a sample of voters on a large number of variables—age, years of education, income, etc. You use a computer to calculate correlation coefficients among 12 variables (66 correlation coefficients). You find that three of these are significant at the 5% level. Do your results support the theory that these three pairs of variables are correlated in the population from which the sample was drawn? Why? _____

_ _

No. When you calculate 66 correlation coefficients you must expect 5% of them (3.3 on the average) to be significant at the 5% level by chance.

PREDICTION

Often we wish to use one measure to predict another measure; for example, we may use aptitude tests to predict academic performance or job performance. We may use advertising volume to predict the sales of a product. We may use rainfall to predict the growth rate of crops. The technique for developing such predictions is called regression analysis.

48. Consider the scattergram below. It represents the relation between a test of finger dexterity and productivity on an assembly job.

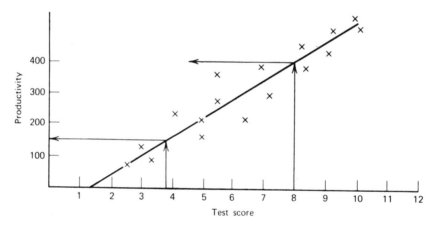

The diagonal line is the straight line that best represents the relationship between the two measures. You can use this line to obtain the best prediction of productivity for any given test score; for example, if a person scores 3.75 on the test, our best prediction of his productivity is 150 units per day. If a person scores 8.0, our best prediction of his productivity is_____.

_ _

400 units per day

49. The line on the scattergram is called the regression line of y on x or, in this case, the _____ of _____ on _____.

– –

regression line
productivity
test scores

50. The regression line of y on x indicates the best prediction of y for every value of x on the basis of the sample used. If there is no relationship between the two measures, the regression line does not add to your knowledge. If there is no relationship between y and x in your sample (if $r = 0$), the best prediction of y is always $\bar{y}$, no matter what value x has; for example, try to draw in the regression line of y on x in the scattergram below. You want to draw a line that shows that $\bar{y}$ is the best prediction of y for every value of x.

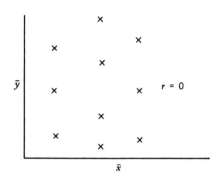

– –

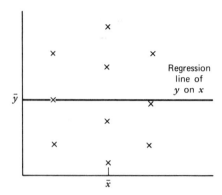

If you use the regression line illustrated above to predict y, your prediction will always be $\bar{y}$, no matter what value of x you start from.

51. When we use a regression line, we assume that the relationship between the two measures is linear. Would a straight regression line give good predictions if the relation between the two measures were strongly curvilinear?

_ _

No

52. It is important to draw a scattergram to make sure that the relationship between the two measures is linear, but it is not necessary to read from the scattergram to make predictions.

The mathematical equivalent of reading the scattergram is accomplished by using the following formula:

$$y = \bar{y} + b(x - \bar{x})$$
$$b = r\frac{s_y}{s_x} \text{ or } \frac{n\Sigma(xy) - (\Sigma x)(\Sigma y)}{n\Sigma x^2 - (\Sigma x)^2}$$

This formula gives the best possible prediction of y for any given value of x. To use the formula you must find the values of $\bar{x}$, $\bar{y}$, and b on the basis of your sample data and then plug in whatever value of x you are interested in.

Let us apply this procedure. The following data represents advertising expenditures (in thousands of dollars) and sales (in tens of thousands of dollars) for a department store. Your problem is to predict the sales that will result from spending $1600 on advertising.

Advertising Expense	Sales
(×$1000)	(×$10,000)
5	4
6	3
1	2
4	6
2	3

These numbers are the same as those used in frame 33, so you can save yourself some arithmetic by looking back at your computations for that frame. First, what are $\bar{x}$ and $\bar{y}$?

$\bar{x} = $ _____

$\bar{y} = $ _____

- -

$\bar{x} = 3.6$
$\bar{y} = 3.6$

53. Now compute b. Refer to the formula above and to your work on frame 33. You have already computed most of the necessary terms.

$b = $ _____ .

$$b = \frac{n\Sigma\,(xy) - (\Sigma x)(\Sigma y)}{n\Sigma x^2 - (\Sigma x)^2} = \frac{5(70) - (18)(18)}{5(82) - (18)^2} = 0.302$$

54. You want to predict the sales resulting from $1600 of advertising. What value of x will you use? (Remember the data is given in thousands of dollars.)

1.6

55. Use the regression equation $y = \bar{y} + b(x - \bar{x})$ to predict the value of y.

$y = $ _____.

$y = \bar{y} + b(x - \bar{x}) = 3.6 + 0.302(1.6 - 3.6) = 3.6 - 0.604 = 3.0$, or $30,000

56. On the basis of the same sample, what sales would you predict for $4600 of advertising?

$y = \bar{y} + b(x - \bar{x}) = 3.6 + 0.302(4.6 - 3.6) = 3.6 + 0.302 = 3.9$ or $39,000

57. When you use a regression equation to predict values of a measure, you are subject to the danger that the relationship between the measures in your sample is entirely or partly due to chance. By making certain complex assumptions about the distribution of y it is possible to establish confidence intervals for your predictions. These procedures are beyond the scope of this book. Another useful check on regression equations is cross validation. You can use the regression equation developed with one sample to predict values of y for a new sample and then check the accuracy of the predictions; for example, a researcher is developing a method based on an auditory discrimination test, of predicting success in learning foreign languages. On the basis of his sample he computes

$\overline{x} = 10$ $\overline{y} = 50$ $b = 5.0$
(discrimination   (language test)
test)

Write the appropriate regression equation for predicting y.

$y = $ _____.

_ _

$y = \overline{y} + b(x - \overline{x}) = 50 + 5.0(x - 10)$

58. To cross-validate his regression equation he will use (a new sample/the same sample).

_ _

a new sample

59. In his new sample the researcher obtains discrimination test scores of 5, 6, 8, 12, and 15. What language learning scores will he predict? Apply his regression equation.

x	y(predicted)
5	
6	
8	
12	
15	

- -

x	y (predicted)
5	25
6	30
8	40
12	60
15	75

60. The actual scores he finds are as follows:

x	y (predicted)	y (actual)
5	25	23
6	30	35
8	40	41
12	60	58
15	75	75

His predictions   appear to be relatively (accurate/inaccurate).

- -

accurate

61.   Describe cross-validation briefly in your own words.

_ _

Your answer should have included these points:

(a)   Obtain new sample.
(b)   Use regression equation to predict scores.
(c)   Compare predicted scores with actual scores.

REVIEW PROBLEMS

If you have successfully completed this chapter, you can now statistically describe and test the relationship between two measures. You can:

- draw a scattergram;
- compute the correlation coefficient and test the null hypothesis that $r = 0$;
- use a regression equation to predict one measure on the basis of another, correlated measure.

Now try these review problems. Table I on pp. 253–256 lists any formulas you may need for reference.

1.   Draw a scattergram for the following data.

x 1, 2, 3, 5, 7, 8, 10, 11
y 1, 2, 4, 5, 4, 3, 1, 1

2. Describe the relation between x and y in the data for problem 1. Is a correlation coefficient an appropriate measure of the strength of the relationship?

3. Compute r for the following data:

x 2, 3, 3, 4, 5, 5, 6
y 7, 6, 5, 4, 3, 2, 1

4. For the data in problem 3 outline an appropriate statistical test for the null hypothesis that there is no correlation between the populations of x and y scores. Use a 1% significance level. Is the r obtained in problem 3 significant?

5. Use the data for problem 3 to establish a regression equation to predict y. If $x = 5$, what is your best prediction for y?

Answers

To review a problem, study the frames indicated after the answer.

1.

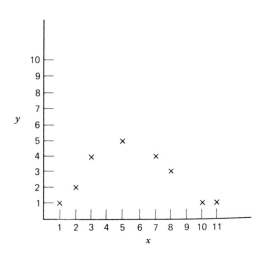

See frames 1 to 4.

2. The relationship is curvilinear; therefore r is not an appropriate measure of the strength of the relationship. *See frames 6 to 11.*

3.

x	x	y	y	xy
2	4	7	49	14
3	9	6	36	18
3	9	5	25	15
4	16	4	16	16
5	25	3	9	15
5	25	2	4	10
6	36	1	1	6
28	124	28	140	94

$$r = \frac{n\Sigma(xy) - (\Sigma x)(\Sigma y)}{\sqrt{[n\Sigma x^2 - (\Sigma x)^2][n\Sigma y^2 - (\Sigma y)^2]}} = \frac{7(94) - (28)(28)}{\sqrt{[7(124) - (784)][7(140) - (784)]}}$$

$$= \frac{-126}{\sqrt{16464}} = \frac{-126}{128} = -0.98$$

See frames 32 to 33.

4. Null hypothesis $\rho = 0$
   Alternative $\rho \neq 0$
   Significance level $\alpha = 0.01$
   Critical region $r \geq +0.875$ or $r \leq -0.875$

   Since $r = -0.98$, the null hypothesis is rejected. *See frames 34 to 45.*

5. $\overline{x} = 4$
   $\overline{y} = 4$

$$b = \frac{n\Sigma(xy) - (\Sigma x)(\Sigma y)}{n\Sigma x^2 - (\Sigma x)^2} = \frac{7(94) - (28)(28)}{7(124) - (28)^2} = \frac{-126}{84} = -1.5$$

$y = \overline{y} + b(x - \overline{x}) = 4 - 1.5(x - 4)$
$y = 4 - 1.5(5 - 4) = 2.5$ if $x = 5$.

See frames 48 to 59.

CHAPTER EIGHT
A Test of Distributions

Many kinds of data come in the form of counts or categories. For example think of the sets at the start of Chapter 1, "blue eyes, brown eyes, green eyes...", or "station wagon, sports car...". Using a sampling distribution called chi square (χ^2), it is possible to determine the probability that any given sample was drawn from a population with a given distribution.

In using chi square for formal hypothesis testing, the important thing is to be able to come up with an appropriate population model for your null hypothesis. With a little ingenuity, it is possible to use chi square tests in many cases where other kinds of statistical tests are of doubtful applicability, for example when measurements do not have normal distributions or equal variances.

When you have completed this chapter, you will be able to:

- perform a χ^2 test with data on a single variable;
- perform χ^2 tests for data on two variables;
- use χ^2 tests as a substitute for analysis of variance or t tests when appropriate.

THE CHI SQUARE TEST OF A DISTRIBUTION

The χ^2 distribution (Greek letter *chi*) is a theoretical sampling distribution that allows you to test the assumption that a sample was drawn from a population with a given distribution. It allows you to compare a sample distribution with a population distribution derived from a theory or null hypothesis and decide whether the sample could reasonably be a random sample from that population.

1. On the basis of genetic theory you predict that a particular population of guinea pigs should be 40% brown, 40% spotted, and 20% white. What distribution of colors would you expect in a sample of 50 guinea pigs? Complete the table below:

	Brown	Spotted	White
Predicted number of guinea pigs			

- -

	Brown	Spotted	White
Predicted number of guinea pigs	20	20	10

2. In a sample of 50 guinea pigs you find 30 brown, 15 spotted, and 5 white. This observation is compared with your theoretical prediction in the following table:

	Brown	Spotted	White
Predicted	20	20	10
Observed	30	15	5

Can the differences from what you predicted be accounted for by chance or will you have to reject the theory on which your predictions are based? To find out you can compute a value of χ^2.

The formula for χ^2 is

$$\chi^2 = \Sigma\left[\frac{(f - F)^2}{F}\right]$$

In this formula F is the predicted frequency for a given category (or "cell") and f is the observed frequency. For our example we compute as follows:

f	F	$(f - F)$	$(f - F)^2$	$(f - F)^2/F$
30	20	+10	100	5
15	20	−5	25	1.25
5	10	−5	25	2.50

$$\Sigma\frac{(f - F)^2}{F} = \overline{8.75}$$

What is the value of χ^2?

- -

$\chi^2 = 8.75$

3. If the differences between the theoretically predicted frequencies (F) and the observed frequencies (f) tend to be large, χ^2 will be (large/small).

- -

large

4. What would a small value of χ^2 indicate?

_ _

The observed frequencies tend to be close to the theoretical predictions; $(f - F)^2$ tends to be small.

5. Look at the formula for χ^2. Is a negative value of χ^2 possible?

_ _

No, $(f - F)^2$ is always positive, so χ^2 is always positive.

6. To find the probability of obtaining by chance a value of χ^2 as large as 8.75 you can refer to the table on page 276. Look at the table now. What additional information do you need to use the table?

_ _

You need to know the df, the degrees of freedom.

7. In this case the degrees of freedom for χ^2 are the number of categories minus 1. What degrees of freedom will you use for this problem?

_ _

2

8. What is the probability of an χ^2 as large as 8.75?

– –

Less than 0.025 (between 0.025 and 0.01)

9. You believe that people are most likely to attempt suicide on a weekend. To test this theory, you have obtained reports of attempted suicides from the police. In the past two years, 147 attempts were reported, distributed as follows:

Day	Sun.	Mon.	Tues.	Wed.	Thurs.	Fri.	Sat.	Total
Number of attempts	32	10	13	13	4	40	35	147

An appropriate null hypothesis for a statistical test would be that equal numbers of attempts are made on all days of the week. Use this theory to compute predicted frequencies for the following table. Divide the total equally among the groups.

	Sun.	Mon.	Tues.	Wed.	Thurs.	Fri.	Sat.
Predicted							

– –

	Sun.	Mon.	Tues.	Wed.	Thurs.	Fri.	Sat.
Predicted	21	21	21	21	21	21	21

(Since there are seven categories, the predicted number for each category would be 1/7 of the total.)

10. What are the degrees of freedom for χ^2 in this case?

_ _

df = 6

11. Use the table to establish a critical region for χ^2 at a 1% significance level.

_ _

$\chi^2 \geq 16.81$

12. Compute χ^2. Can you reject the null hypothesis?

_ _

f	F	$(f - F)$	$(f - F)^2$	$(f - F)^2/F$
32	21	11	121	5.8
10	21	− 11	121	5.8
13	21	− 8	64	3.0
13	21	− 8	64	3.0
4	21	− 17	289	13.8
40	21	19	361	17.2
35	21	14	196	9.3
				57.9

$\chi^2 = 57.9$

The null hypothesis can be rejected.

13. There is one important fact to note about this statistical test. The order of the categories has no effect on the value of χ^2. Only the amount of the differences matters. As a result, you would have obtained the same value of χ^2 if your distribution had been the following:

Day	Sun.	Mon.	Tues.	Wed.	Thurs.	Fri.	Sat.
Number of reports	13	13	35	40	32	10	4

Would these results have supported your theory? Why?

- -

No. These data indicate that suicide attempts are made in mid-week.

14. When you use a χ^2 test to reject a null hypothesis, you must always look back at the data to make sure that they support your alternative. You have taken a census of the insect population of your rose garden. On the basis of several large samples you conclude that the insect population is distributed as follows:

Ladybugs   20%
Inch worms   20%
Weevils   30%
Aphids   25%
Brown spiders 5%

Now you treat your garden with an insecticide that is supposed to control the undesirable weevils, aphids, and brown spiders without affecting ladybugs or inch worms.

To check the effect of the insecticide you collect 150 insects at random. Your sample is composed as follows:

Ladybugs   25
Inch worms   45
Weevils   45
Aphids   25
Brown spiders 10

Use the census to determine the predicted frequencies. Compute χ^2 and answer these two questions:

1. Has the distribution of insects changed significantly (at the 5% level)?
2. Has the insecticide had the intended effect?

f	F	$(f - F)$	$(f - F)^2$	$(f - F)^2/F$
25	30	5	25	.83
45	30	15	225	7.50
45	45	0	0	0
25	37.5	12.5	156.25	4.17
10	7.5	2.5	6.25	0.83
				13.33

$X^2 = 13.33$
df $= 4$
Critical region $\chi^2 \geq 9.49$

(a)   The distribution of insects has changed significantly.
(b)   The insecticide does not have the intended effect. In theory the proportions of ladybugs and inch worms should increase as the proportions of weevils, aphids, and brown spiders decrease. Instead, weevils were unaffected, ladybugs decreased, and brown spiders increased. Only inch worms and aphids changed in the predicted direction.

CHI SQUARE TEST WITH TWO VARIABLES

Chi square can also be used to test hypotheses about distributions based on two variables. To do so we must establish a set of categories based on the two variables and predict the frequency of observations in each category.

15.   A researcher believes that people tend to choose mates with the same hair color. He can test this theory by constructing a distribution based on two variables: the husband's hair color and the wife's hair color. His table will look like this:

Wife	Husband			
	Red	Blond	Black	Brown
Red				
Blonde				
Black				
Brown				

How many cells or categories are in this table?

_ _

16. Each possible combination of husband and wife is a category.

16. The researcher's first problem is to calculate a predicted frequency for each cell. To do this he must consider the numbers of men and women in each category. He has observed, let us say, 500 couples. The numbers of men and women in each category are indicated on the edges of the following table. In addition, the percentage of men in each category is listed at the bottom for reasons you will understand shortly.

Wife	Red	Blond	Black	Brown	Total
Red					50
Blonde					150
Black					150
Brown					150
Total	50	100	150	200	500
%	10%	20%	30%	40%	100%

(Husband across the top columns: Red, Blond, Black, Brown, Total)

The totals along the edge of the table are called marginals. Look at the marginals to find out how many of the wives are blondes.

_ _

150

17. How many of the husbands have brown hair? What percent?

_ _

200
40%

18. Forty percent of all the husbands have brown hair. If there is no relation between the husband's hair color and the wife's hair color, approximately 40% of the women in each category should have husbands with brown hair; for example, of the 150 blondes, how many should have husbands with brown hair?

_ _

$0.40 \times 150 = 60$

19. Ten percent of all husbands have red hair. How many of the 50 red-headed wives should have red-headed husbands?

- -

$0.10 \times 50 = 5$

20. How many of the brown-haired wives should have blond husbands?

- -

$0.20 \times 150 = 30$

21. Apply this method to fill in predicted frequencies in the table.

Wife	Red	Blond	Black	Brown	Total	%
Red	$F =$	$F =$	$F =$	$F =$	50	10%
Blonde	$F =$	$F =$	$F =$	$F =$	150	30%
Black	$F =$	$F =$	$F =$	$F =$	150	30%
Brown	$F =$	$F =$	$F =$	$F =$	150	30%
Total	50	100	150	200	500	
%						

- -

5	10	15	20
15	30	45	60
15	30	45	60
15	30	45	60

22. Degrees of freedom in this case depend on the number of categories in each direction. They are $(c - 1)(r - 1)$, the number of columns minus one times the number of rows minus one. Columns are the categories of husbands. How many columns are there in this problem?

- -

4

23. The rows in this problem are the categories for _____.

_ _

wives

24. In this problem df $= (c - 1)(r - 1) = $ _____.

_ _

$(4 - 1)(4 - 1) = 9$

25. We have added the observed frequencies to the table. Compute x^2 and answer two questions.

(a) Is χ^2 significant at the 1% level?
(b) Does the data support the researcher's theory that people tend to select mates with the same hair color?

Wife   Husband

		Red	Blond	Black	Brown	Total
Red	F	5	10	15	20	50
	f	10	10	10	20	
Blonde	F	15	30	45	60	150
	f	10	40	50	50	
Black	F	15	30	45	60	150
	f	13	25	60	52	
Brown	F	15	30	45	60	150
	f	17	25	30	78	
Total		50	100	150	200	500

f	F	$(f - F)$	$(f - F)^2$	$(f - F)^2/F$
10	5	5	25	5.00
10	15	− 5	25	1.67
13	15	− 2	4	.27
17	15	2	4	.27
10	10	0	0	0
40	30	10	100	3.33
25	30	− 5	25	.83
25	30	− 5	25	.83
10	15	− 5	25	1.67
50	45	5	25	.56
60	45	15	225	5.00
30	45	− 15	225	5.00
20	20	0	0	0
50	60	− 10	100	1.67
52	60	− 8	64	1.07
78	60	18	324	5.40

$$\chi^2 = 32.57$$

(a) χ^2 is significant; df $= 9$. The critical region is $\chi^2 \geq 21.67$.

(b) The data supports the theory because the observed frequency of matched husband and wife couples is greater than predicted in all cases, whereas most of the other observed frequencies are equal to or less than the prediction.

26. As a rule, to use the χ^2 test the predicted frequency for each cell should be at least 5.

There are exceptions to this rule but they are beyond the scope of this book.

Could you have performed the χ^2 test in frame 18 if you had a sample of 100 couples?

– –

No. Many of the predicted frequencies would be less than 5.

27. You are investigating the sensitivity of goldfish to industrial pollutants in their water. Some fish appear to be very sensitive. They turn white and die instantly. Others are moderately sensitive. They show signs of distress, but do not die, and many show no reaction at all. Since you are working with both black and red goldfish, you would like to know if there is evidence of a difference in sensitivity for the two colors. The following tables classifies a group of goldfish according to color and sensitivity. Use the marginals to calculate predicted frequencies on the assumption that there is no relation between color and sensitivity and fill them in in the table.

		Sensitivity				
Color		High	Medium	Low	Total	%
Red	F					
	f	4	4	22	30	60%
Black	F					
	f	6	6	8	20	40%
Total		10	10	30	50	

- -

		Sensitivity				
Color		High	Medium	Low	Total	%
Red	F	6	6	18		
	f	4	4	22	30	60%
Black	F	4	4	12		
	f	6	6	8	20	40%
Total		10	10	30	50	

28. Can you use χ^2 test on this data as it is? Why?

- -

No. Two of the predicted frequencies are less than 5.

29. When the predicted frequencies are less than 5, it is sometimes possible to combine categories so that the predicted frequencies are large enough. In this case you could logically combine two of the sensitivity categories in order to obtain predicted frequencies larger than 5. Prepare a new table that does this. Your table will now have only 4 cells.

	Color		Sensitivity		Total	%
			High or Moderate	Low		
Red	F		12	18		
	f		8	22	30	60%
Black	F		8	12		
	f		12	8	20	40%
Total			20	30	50	

30. What are the degrees of freedom for this new table?

- -

$1, (2 - 1)(2 - 1)$

31. When you use a computer to calculate chi square you must still decide on the predicted frequencies and often compute them by hand.

Study the following printout.

CHI SQUARE TEST

CATEGORY	CASES OBSERVED	CASES EXPECTED	RESIDUAL
1	29	22.00	7.00
2	19	20.00	−1.00
3	18	18.00	0.00
4	25	18.00	7.00
5	17	18.00	−1.00
6	10	18.00	−8.00
7	15	16.00	−1.00
8	11	14.00	−3.00
TOTAL	144		

CHI SQUARE	DF	SIGNIFICANCE
9.316	7	0.231

What label corresponds to f in our formula? _____
What label corresponds to F? _____ How often would
you expect to obtain the results in the printout by chance? _____

- -

f = CASES OBSERVED
F = CASES EXPECTED
p = 0.231 You would expect to obtain a value of χ^2 this great by chance
about 23% of the time.

WHEN TO USE CHI SQUARE

The t test for comparing two means and the analysis of variance both involve assumptions that the populations are normally distributed and have equal variances. A χ^2 test can often be used when the data does not meet these assumptions. Instead of using the measurements to calculate a mean, you simply use them to categorize individual observations; for example, as "high," "medium," and "low" or above and below the mean. The choice of categories should be guided by your theory and by the possibility of constructing an appropriate set of theoretical frequencies.

32. Here is a problem you considered earlier. The following is a summary of the adjustment scores of mental patients treated with three different drugs.

Group 1	Group 2	Group 3
$\overline{x} = 5$	$\overline{x} = 10$	$\overline{x} = 4$
$s^2 = 4$	$s^2 = 23$	$s^2 = 5$
$n = 10$	$n = 33$	$n = 45$

Analysis of variance is not appropriate because the variances of the three groups are unequal. What categories could you use to set up an χ^2 test?

- -

Group 1, Group 2, Group 3, high adjustment score, and low adjustment score.

33. When you can use a t test or analysis of variance, it is better to do so because the power of these tests is greater than the power of a χ^2 test; that is, you are more likely to be able to reject the null hypothesis if your theory is true. You are interested in whether educational background makes a difference in scores on a trouble-shooting test. You have three samples of 25 people each with the following backgrounds: college, high school, and technical institute. The trouble-shooting scores appear to be normally distributed and the sample variances are all quite similar. Could you use χ^2 for examining this data?

- -

Yes, by using the trouble-shooting scores to categorize individuals, for example, as high, medium, and low.

34. Could you use analysis of variance for examining this data?

––––––––––––––––––––––––

Yes

35. Which type of statistical test would you normally choose in this case?

––––––––––––––––––––––––

Analysis of variance

36. You want a quick, informal check on approximately how unusual a distribution is. Either χ^2 or analysis of variance could be used. Which one will you choose?

––––––––––––––––––––––––

We would choose χ^2 because it is usually easier to compute, and precise formal hypothesis testing is not required in this case.

REVIEW PROBLEMS

If you have successfully completed this chapter, you can now use the *chi square* distribution to test hypotheses about the distribution of a population. You can:

- develop an appropriate theoretical population distribution and perform a χ^2 test with frequency data on a single variable;
- develop an appropriate theoretical population distribution and perform a χ^2 test with frequency data on two variables;
- recognize situations in which χ^2 can be used as a substitute for analysis of variance or t tests.

Now try these review problems. Table I on pp. 253–256 lists any formulas you may need for reference.

1. A factory has four machines that produce molded parts. A sample of 500 parts is collected for each machine and the number of defective parts in each sample is determined:

Machine	1	2	3	4
Defects/500	10	25	0	5

Is there a difference between the machines? Outline an appropriate statistical test and, if possible, determine if these data are significant.

2. The situation is the same as in problem 1 but you are measuring the mean diameter of the parts. You suspect that the mean diameters for parts produced on the four machines are not the same. What would be your first choice of a statistical test? What assumptions would be required?

3. The dropout rate among volunteer workers in community development programs varies widely. The theory is advanced that the degree of involvement of volunteers in setting the goals of the program influences the dropout rate. In a new program 27 volunteers are selected to participate in a series of goal-setting workshops as part of their activities, whereas 23 others are simply assigned tasks. At the end of two months the results are as follows:

	Remained in Program	Dropped Out	Total
Workshop group	18	9	27
No workshop group	10	13	23
Total	28	22	

Are these results significant? Outline an appropriate statistical test and compute the necessary statistic.

4. As part of an attitude survey, a sample of men and women is asked to rate a number of statements on a scale of 1 to 5, according to whether they agree or disagree. The following are the results for one of the statements.

	Agree Strongly 1	2	3	4	Disagree Strongly 5	Total
Women	3	13	10	16	7	50
Men	2	12	26	10	1	50
Total	5	25	36	26	8	100

Use χ^2 to determine whether there is a significant difference between the answers of men and women.

Answers

To review a problem, study the frames indicated after the answer.

1. You can use χ^2 for this problem. The theoretical distribution would call for an equal number of defects from each machine.

Machine	1	2	3	4	Total
F	10	10	10	10	40
f	10	25	0	5	40

F is greater than 5 for all cells.

f	F	$(f - F)$	$(f - F)^2$	$(f - F)^2/F$
10	10	0	0	0
25	10	15	225	22.5
0	10	− 10	100	10.0
5	10	− 5	25	2.5
				$\chi^2 = 35.0$
				df $= 3$

For $\alpha = 0.05$ the critical region is $\chi^2 \geq 7.81$.
For $\alpha = 0.01$ the critical region is $\chi^2 \geq 11.34$.
There is a significant difference among machines. *See frames 1 to 11.*

2. Your first choice should be analysis of variance, since you are dealing with measurements rather than count data. This test will give you the greatest power and the best chance of rejecting the null hypothesis if your theory is true. To use analysis of variance you must assume that the populations of measurements **for all machines are approximately normally distributed and have approxi-mately equal variances.** *See frames 32 to 36 and Chapter 6, frames 69 to 74.*

3. χ^2 is an appropriate test. The null hypothesis is that the workshop has no effect on the dropout rate and that equal numbers drop out from both groups. The alternative is that different numbers will drop out. The critical region for $\alpha = 0.05$ is $\chi^2 \geq 3.84$ (df = 1).

f	F	$(f - F)$	$(f - F)^2$	$(f - F)^2/F$
18	15.21	2.88	8.29	0.548
10	12.88	−2.88	8.29	0.644
9	11.88	−2.88	8.29	0.698
13	10.12	2.88	8.29	0.819
				$\chi^2 = 2.709$

The results are not significant.

4. Half of the sample is women and half men. If there is no difference in their responses, you would expect equal numbers of men and women in each category of response. The predicted frequencies on this basis are

	1	2	3	4	5
Women	2.5	12.5	18	13	4
Men	2.5	12.5	18	13	4

Because four of the predicted frequencies are less than 5, you must combine categories. By combining categories 1 and 2 and 4 and 5 you may compute χ^2.

		1 and 2	3	4 and 5
Women	F	15	18	17
	f	16	10	23
Men	F	15	18	17
	f	14	26	11

f	F	$(f - F)$	$(f - F)^2$	$(f - F)^2/F$
16	15	1	1	0.067
14	15	−1	1	0.067
10	18	−8	64	3.556
26	18	8	64	3.556
23	17	6	36	2.118
11	17	−6	36	2.118
				$\chi^2 = 11.482$

For $\alpha = 0.01$ the critical region is $\chi^2 \geq 9.210$ (df = 2). The difference is signifi-cant. *See frames 26 to 30.*

CHAPTER NINE

The Combined Effects of Two Variables

Analysis of variance can be extended to study the combined effect of two experimental treatments on a measure. This method of analysis allows for the design of experiments that are particularly efficient. That is, you can test a large number of hypotheses from relatively few experimental observations. Despite what you may feel at the end of this book, experimental observations are usually more work than statistical analysis. As a result, two-way analysis of variance is a very popular technique for experimenters.

When you have completed this chapter, you will be able to perform and interpret a simple two-way analysis of variance.

1. Often a researcher is interested in the combined effects of two variables on a third measure; for example, consider this problem. A researcher is trying to determine the conditions that will produce the greatest yield from melon vines. He suspects that both the amount of fertilizer used and the amount of water the plants receive will influence the number of melons on a vine. In this case, the two variables whose effect is under study are _____ and _____.

- -

amount of fertilizer
amount of water

2. The measure under study is _____.

- -

number of melons per vine

3. To study the effect of the amount of water alone the researcher might cultivate 20 melon vines with heavy watering and 20 with light watering and count the number of melons per vine in each group. What sort of statistical test could he use. What assumptions would be involved?

— —

He could use a t test for the difference between two means, assuming that the number of melons per vine is approximately normally distributed and that the variances would be approximately equal under the two conditions. If he could not make these assumptions, he could use a χ^2 test by counting the number of vines under each condition that produced above or below average numbers of melons.

4. To study the effect of amount of fertilizer alone he might cultivate three groups of vines, one with low fertilizer, one with a moderate amount, and one with a high amount. What sort of statistical test could he use for this experiment? What assumptions are involved?

— —

He could use analysis of variance, assuming approximately normal distributions and equal variances, or he could use a χ^2 test.

5. Suppose that there is an interaction between the effects of the water and fertilizer. Perhaps fertilizer is effective only if there is also heavy watering or perhaps fertilizer helps to compensate for the effects of light watering. Will a study of watering alone or of fertilizing alone detect this interaction?

— —

No

6. In this kind of situation it is possible, in effect, to perform both studies at once and study the results with a two-way analysis of variance. To do this the experimenter will set up six experimental groups as follows:

Water	Fertilizer		
	Light	Medium	Heavy
Heavy			
Light			

If he selects 10 plants at random for each group, how many plants will he have in all?

_ _

60

7. In all, how many plants will receive heavy watering?

_ _

30

8. In all, how many plants will receive light fertilizing?

_ _

20

9. Let us establish some terminology. We call each group of 10 a cell; for example, the plants with heavy watering and medium fertilizer make up a cell. We call each level of watering a row; for example, all plants with light watering make up a row. We call each level of fertilizing a column; for example, all plants with light fertilizing make up a column.

All plants with medium fertilizing would be called a _____.

_ _

column

10. All plants with medium fertilizing and light watering are called a _____.

_____

cell

11. All plants with heavy watering are called a _____.

_____

row

12. Here are the results of the experiment:

Fertilizer

Water	Light		Medium		Heavy		Totals	
	Number of Melons	x^2	Number of Melons	x^2	Number of Melons	x^2	x	x^2
Heavy	1	1	4	16	6	36		
	2	4	5	25	7	49		
	2	4	5	25	7	49		
	3	9	6	36	8	64		
	3	9	6	36	8	64		
	3	9	6	36	8	64		
	3	9	6	36	8	64		
	4	16	7	49	9	81		
	4	16	7	49	9	81		
	5	25	8	64	10	100		
	30	102	60	372	80	652	170	1126
Light	5	25	3	9	0	0		
	6	36	4	16	1	1		
	6	36	4	16	1	1		
	7	49	5	25	2	4		
	7	49	5	25	2	4		
	7	49	5	25	2	4		
	7	49	5	25	2	4		
	8	64	6	36	3	9		
	8	64	6	36	3	9		
	9	81	7	49	4	16		
	70	502	50	262	20	52	140	816
Totals	100	604	110	634	100	704		
Grand total							310	1942

As you can see, we have already computed Σx or Σx^2 for each cell, row, and column and for the total group. What is Σx for the heavy row?

_ _

170

13. What is Σx^2 for the light column?

_ _

604

14. What is Σx_T?

_ _

310

15. As you know, analysis of variance allows us to estimate the variance of the population on the basis of the differences between group means and to compare this variance estimate with an estimate based on the individual differences within groups. We can group the data for a variance estimate based on group means in several ways. We can look at six individual cell means, at two row means, or at three column means. How many different variance estimates can we obtain from these groupings?

_ _

Three. The variance of the individual cell means can be used to estimate population variance; so can the variance of row means and the variance of column means.

16. Here are the formulas you will use for a two-way analysis of variance. Some of them you have already used for the one-way analysis of variance. Some of them are new. Which ones are new? Place a check mark next to the new ones.

Total sum of squares

$$\Sigma x_T^2 - \frac{(\Sigma x_T)^2}{N}$$
df $= N - 1$

Between groups sum of squares

$$\frac{(\Sigma x_1)^2}{n} + \frac{(\Sigma x_2)^2}{n} + \cdots - \frac{(\Sigma x_T)^2}{N}$$
df $= rc - 1$

Between rows sum of squares

$$\frac{(\Sigma x_{row1})^2}{nc} + \frac{(\Sigma x_{row2})^2}{nc} + \cdots - \frac{(\Sigma x_T)^2}{N}$$

df $= r - 1$

Between columns sum of squares

$$\frac{(\Sigma x_{col1})^2}{nr} + \frac{(\Sigma x_{col2})^2}{nr} + \cdots - \frac{(\Sigma x_T)^2}{N}$$
df $= c - 1$

Interaction sum of squares

between groups sum of squares—(between rows sum of squares + between columns sum of squares) or

$$\frac{(\Sigma x_1)^2}{n} + \frac{(\Sigma x_2)^2}{n} + \cdots - \frac{(\Sigma x_{row1})^2}{nc} - \frac{(\Sigma x_{row2})^2}{nc} - \cdots - \frac{(\Sigma x_{col1})^2}{nr} - \frac{(\Sigma x_{col2})^2}{nr} - \cdots + \frac{(\Sigma x_T)^2}{N}$$

df $= (r - 1)(c - 1)$

Within groups sum of squares

total sum of squares—between groups sum of squares or

$$\left[\Sigma x_1^2 - \frac{(\Sigma x_1)^2}{n}\right] + \left[\Sigma x_2^2 - \frac{(\Sigma x_2)^2}{n}\right] + \cdots$$
df $= N - rc$

- -

The new ones are:

Between rows sum of squares
Between columns sum of squares
Interaction sum of squares

17. The between groups sum of squares is based on the cells. It tells you exactly the same thing it would in a one-way analysis of variance. If you divide the between groups variance by the within groups variance and obtain a significant value of F, what does it tell you? Think of the null hypothesis.

– –

It tells you that all the cell means are not equal.

18. Will a significant between groups variance tell you anything about whether water or fertilizer, or both, is the cause of the differences?

– –

No. You know only that at least one cell mean is significantly different from the others.

19. Look closely at the formula for between groups sum of squares. It is identical to the one you used for one-way analysis of variance. In the formula for degrees of freedom r stands for the number of rows and c, for the number of columns. In our problem what are the degrees of freedom for the between groups variance?

– –

$(2 \times 3) - 1 = 5$

20. Now look at the formulas for between rows sum of squares and between columns sum of squares. They are the same as the between groups formula, except that they are based on rows and columns instead of the individual cells. In our problem what value corresponds to Σx_{row_1}?

– –

170

21. What value corresponds to Σx_{col_1}?

– –

100

22. In the formula for between rows sum of squares n stands for the number of measurements in a cell and c, for the number of columns. The number of measurements in a row is nc. In our problem this is _____.

_ _

30

23. If we divide the between rows variance estimate by the within groups variance estimate and obtain a significant F, we can conclude that _____ makes a difference in the mean number of melons on a plant.

_ _

water

24. If the between columns variance estimate is significant, we can conclude that _____ makes a difference.

_ _

fertilizer

25. The between groups sum of squares will usually be larger than the sum of the between rows and between columns sums of squares; that is, part of the variance between cells will not be explained by either the difference between rows or the difference between columns. This remaining part of the difference between cells must be due to the combined effect of the two variables: that is, interaction.

Suppose that fertilizer improves the yield of vines under most conditions but heavy watering combined with a medium application of fertilizer washes away the fertilizer so that it has no effect. In this case you would expect two significant variance estimates; _____ and _____.

_ _

between columns
interaction

26. The results of a two-way analysis of variance for the melon problem are summarized below:

	Sum of squares	df	Variance Estimate	F
Total	340.33	59		
Between rows	15.00	1	15.00	11.28
Between columns	3.33	2	1.67	1.26
Interaction	250.00	2	125.00	93.99
Within groups	72.00	54	1.33	

(The between groups sum of squares is 268.33, but this figure is not usually included in a summary because the rows, columns, and interaction sums of squares give all the information.)

Are any of these F ratios significant? If so, which ones?

- -

Between rows and interaction are both significant. Note that the F table has no entry corresponding to df $= 54$. The values for df $= 50$ and df $= 55$ are close enough to ensure that the results are significant.

27. To interpret the results of a two-way analysis of variance it is helpful to examine the cell means and compare them with a mean for each row and column and a grand total mean.

Water	Fertilizer			
	Light	Medium	Heavy	All conditions
Heavy	3.0	6.0	8.0	5.7
Light	7.0	5.0	2.0	4.7
All conditions	5.0	5.5	5.0	5.2 ◄

Grand total

What combination produces the greatest yields?

- -

Heavy watering and heavy fertilizer

28. What combination produces the poorest yields?

- -

Light watering and heavy fertilizer

29. You are planning to raise melons in a situation in which you cannot control or predict the amount of watering. There is no irrigation and rainfall is unpredictably light or heavy. A low yield would be a disaster. What level of fertilizer would you use on the basis of the data in the experiment?

_ _

Medium. It ensures a relatively high yield no matter what the level of watering. Either heavy or light carries a risk of lower yields, although both also bring a chance of higher yields.

30. Of course, the labor of computing an analysis of variance can be handled by a computer with great ease. A printout of the results of the melon problem would look like this:

SOURCE OF VARIATION	SUM OF SQUARES	DF	MEAN SQUARE	F	SIGNIFICANCE
MAIN EFFECTS					
WATER	15.00	1	15.00	11.28	0.000
FERTILIZER	3.33	2	1.67	1.26	0.691
2-WAY INTERACTIONS					
WATER FERTILIZER	250.00	2	125.00	93.99	0.000
RESIDUAL	72.00	54	1.33		
TOTAL	340.33	59	5.77		

CELL MEANS

	WATER	
FERTILIZER	1	2
1	3.0 (10)	7.0 (10)
2	6.0 (10)	5.0 (10)
3	8.0 (10)	2.0 (10)

In this particular printout the within groups result is called by a different name. What is the within groups result called here? _____

_ _

RESIDUAL

31. The two-way analysis of variance involves the same assumptions as other forms of analysis of variance. What are they?

– –

The populations involved are approximately normally distributed and have approximately equal variances.

32. You expect one of your experimental treatments to have a dramatic effect on the variability of the data. Should you plan a two-way analysis of variance for your statistical test?

– –

No

33. You are studying measures that in your general experience are bimodally distributed. Should you plan to use analysis of variance for a statistical test?

– –

No. In this case a χ^2 test would probably be appropriate, since the data seem to fall into two groups.

34. A chemist is studying the production of an organic compound in a process that involves a catalyst. She believes that the amount of the compound produced in a given processing time will vary according to the catalyst used and according to the temperature but that the relative effectiveness of the different catalysts is not influenced by the temperature. Assume that five catalysts and four temperature levels are to be investigated. The measure to be used will be units of the compound produced in a 5-minute processing run. The chemist will make a total of 200 processing runs.

    (a) Outline the design for an analysis of variance study. What will the rows and columns be and how many measurements will be made for each cell?

    (b) According to the chemist's theory, which F ratios should be significant? Which should not be significant?
        Be sure to consider all F ratios.

– –

(a) Heat

Heat	A	B	C	D	E
1					
2		10 measurements per cell			
3					
4					

Catalyst

(b) Both the row and column variances should be significant, but there should be no significant interaction because this would indicate a situation in which the relative effectiveness of the catalysts is influenced by temperature.

35. A researcher believes that he has developed a test of language-learning ability that will predict a student's ability to learn any language. He proposed to evaluate his test by using it to classify students as high, medium, or low in learning ability, giving them language training and then achievement tests. To make sure that the test works uniformly well for several languages he will train and test students in Spanish, Russian, Swahili, and Japanese. Thus he plans a two-way analysis of variance. In general, a student is trained and tested on only one language, but the researcher ran short of high-ability students. He then trained and tested 10 high-ability students in two languages each. Is this an acceptable procedure?

No. Analysis of variance assumes independent random samples.

36. Assuming that the researcher obtains an appropriate sample and sets up his study so that the different languages are the columns, which F ratios should be significant?

The F ratio for rows (ability level) would be significant. The interaction should *not* be significant, since he expects the test to work equally well for all languages.

37. What would a significant F ratio for columns (languages) mean?

That the achievement tests are not of equal difficulty or that some languages are more difficult to learn than others.

REVIEW PROBLEMS

If you have successfully completed this chapter, you can now perform and interpret a two-way analysis of variance. You can:

● compute F ratios for rows, columns, and interaction;
● examine cell means to interpret the results of two-way analysis of variance;
● recognize situations were a two-way analysis of variance is not appropriate and suggest alternative tests.

Now try these review problems. Table I on pp. 253–256 lists any formulas you may need for reference.

1. A study is conducted of the effects of motivation and method of instruction on achievement in classroom learning. One hundred high school students are assigned at random to four groups. Half the students are promised 50¢ per point on the final exam. The others are paid $5 to participate in the experiment, regardless of how they score. Half the "high motivation" group (the groups paid 50¢ per point) and half the "low motivation" group (the group paid regardless of score) are taught by one instructor who employs an inductive discovery approach. The remaining students are taught by a second instructor, who employs an expository deductive approach. A standardized final exam is administered to all students. The results of the study are as follows

	Inductive	Deductive	Total
High motivation	$\bar{x} = 54$	$\bar{x} = 55$	54.5
	$(s = 5.1)$	$(s = 5.3)$	
Low motivation	$\bar{x} = 56$	$\bar{x} = 35$	45.5
	$(s = 4.9)$	$(s = 6.1)$	
Total	55	45	

An analysis of variance shows no significant effect of motivation or of method of instruction, but the interaction is significant at the 5% level. The researcher concludes on the basis of these results that the inductive method is superior for poorly motivated students, whereas the deductive method is superior for well-motivated students. Does the data support these conclusions? Why?

2. The following are dexterity scores for boys and girls at three age levels. A high score indicates many errors and low dexterity. Complete a two-way analysis of variance and interpret the results. Refer to the formulas on page 253 for guidance.

	Age		
	5–7	7–9	9–11
	5	3	1
	6	4	2
Girls	6	5	2
	6	6	2
	7	7	3
	6	3	1
	7	4	2
Boys	7	5	3
	8	6	4
	8	7	5

3. You conclude from a two-way analysis of variance that age appears to be related to dexterity, but sex is not. What statistical techniques could you use to further describe and analyze this relationship?

4. Summarize briefly the assumptions required for
    (a) a z test for the difference between the means of two samples
    (b) a X^2 test
    (c) a t test for the difference between the means of two samples
    (d) analysis of variance
    (e) a test of the null hypothesis $\rho = 0$

Answers

To review a problem, study the frames indicated after the answer.

1. There are two major difficulties with the researcher's interpretation of his results.
    (a) The cell means do not fit his description of the interaction. In fact, the inter-
    action appears to consist only of the fact that deductive students with low
    motivation do less well than all the other groups.
    (b) Because each method was used by a single instructor, it is impossible to
    separate the method of instruction from his other characteristics—the loudness
    of his voice and the radiance of his smile. You can reject the null hypothesis that
    there is no difference between the groups, but there are numerous plausible
    alternatives. *See frames 25, 34 to 37, and Chapter 4, frames 26 to 31.*

2.

		5–7		7–9		9–11		Totals	
		x	x^2	x	x^2	x	x^2	x	x^2
		5	25	3	9	1	1		
Girls		6	36	4	16	2	4		
		6	36	5	25	2	4		
		6	36	6	36	2	4		
		7	49	7	49	3	9		
		30	182	25	135	10	22	65	339
		6	36	3	9	1	1		
		7	49	4	16	2	4		
Boys		7	49	5	25	3	9		
		8	64	6	36	4	16		
		8	64	7	49	5	25		
		36	262	25	135	15	55	76	452
Totals		66	444	50	270	25	77	141	791

Total sum of squares:

$$\Sigma x_T^2 - \frac{(\Sigma x_T)^2}{N} = 791 - \frac{(141)^2}{30} = 791 - 662.7 = 128.3$$

df $= N - 1 = 29$

Between groups sum of squares:

$$\frac{(\Sigma x_1)^2}{n_1} + \frac{(\Sigma x_2)^2}{n_2} + \frac{(\Sigma x_3)^2}{n_3} + \frac{(\Sigma x_1)^2}{n_1} + \frac{(\Sigma x_5)^2}{n_5} + \frac{(\Sigma x_6)^2}{n_6} - \frac{(\Sigma x_T)^2}{N} =$$

$$= \frac{(30)^2}{5} + \frac{(25)^2}{5} + \frac{(10)^2}{5} + \frac{(36)^2}{5} + \frac{(25)^2}{5} + \frac{(15)^2}{5} - 662.7 = 180.00 + 125.00 +$$

$$+ 20.00 + 259.20 + 125 + 45 - 662.7 = 91.50$$

df $= g - 1 = 5$

Between rows sum of squares

$$\frac{(\Sigma x_{r1})^2}{nc} + \frac{(\Sigma x_{r2})^2}{nc} - \frac{(\Sigma x_T)^2}{N} = \frac{(65)^2}{15} + \frac{(76)^2}{15} - 662.7 = 281.67 + 385.07 - 662.7 =$$

$$= 4.04$$

df $= r - 1 = 1$

Between columns sum of squares

$$\frac{(\Sigma x_{c1})^2}{nr} + \frac{(\Sigma x_{c2})^2}{nr} + \frac{(\Sigma x_{c3})^2}{nr} - \frac{(\Sigma x_T)^2}{N} = \frac{(66)^2}{10} + \frac{(50)^2}{10} + \frac{(25)^2}{10} - 662.7 =$$

$$= 435.6 + 250.0 + 62.5 - 662.7 = 85.40$$

Interaction sum of squares:
Between groups − (between rows + between columns) = 91.50 − (85.40 + + 4.04) = 2.06

Within groups sum of squares:
Total − between groups = 128.30 − 91.50 = 36.80

	Sum of Squares	df	Variance Estimate	F
Total	128.30	29		
Rows (sex)	4.04	1	4.040	2.64
Columns (age)	85.40	2	42.700	27.95
Interaction	2.06	2	1.030	0.67
Within groups	36.80	24	1.533	

The age groupings have a significant effect on scores but sex has not. *See frames 16 to 29.*

3. The techniques of correlation and regression would be appropriate—scattergrams, correlation coefficients, and regression equations. *See Chapter 7.*

4. (a) Both samples are larger than 30.
   See Chapter 5, frames 28 to 34.
   (b) All cells have predicted frequency greater than 5.
   See Chapter 8, frames 26 to 30.
   (c) Both populations are approximately normally distributed and have approximately equal variances.
   See Chapter 5, frames 28 to 34.
   (d) All populations are approximately normally distributed and have approximately equal variances.
   See Chapter 6, frames 69 to 74.
   (e) Both scores are normally distributed; sample is drawn at random.
   See Chapter 7, frames 35 to 45.

TABLE I REFERENCE FORMULAS

Parameters and Statistics

$$\mu = \frac{\Sigma x}{n} \qquad \overline{x} = \frac{\Sigma x}{n}$$

$$\sigma = \sqrt{\frac{\Sigma(x - \mu)^2}{n}} \text{ or } \sqrt{\frac{\Sigma x^2 - (\Sigma x)^2/n}{n}}$$

$$s = \sqrt{\frac{\Sigma(x - \overline{x})^2}{n - 1}} \text{ or } \sqrt{\frac{\Sigma x^2 - (\Sigma x)^2/n}{n - 1}}$$

$$z = \frac{\overline{x} - \mu}{\sigma_{\overline{x}}} \text{ or } \frac{\overline{x} - \mu}{\sigma/\sqrt{n}}$$

$$r = \frac{1}{n - 1} \Sigma \frac{(x - \overline{x})}{s_x} \cdot \frac{(y - \overline{y})}{s_y} \text{ or } \frac{n\Sigma(xy) - (\Sigma x)(\Sigma y)}{\sqrt{[n\Sigma x^2 - (\Sigma x)^2][n\Sigma y^2 - (\Sigma y)^2]}}$$

Central Limit Theorem for Large Samples

$$\mu_{\overline{x}} = \mu$$

$$\sigma_{\overline{x}} = \frac{\sigma}{\sqrt{n}}$$

Regression Analysis

$$y = \overline{y} + b(x - \overline{x})$$

$$b = \frac{n\Sigma(xy) - (\Sigma x)(\Sigma y)}{n\Sigma x^2 - (\Sigma x)^2}$$

Confidence Intervals

$$\overline{x} \pm z_0 \frac{s}{\sqrt{n}}$$

$$\overline{x} \pm t_0 \frac{s}{\sqrt{n}}$$

$$p \pm z_0 \frac{\sqrt{pq}}{\sqrt{n}}$$

Hypothesis Testing

Null Hypothesis $\mu = C$

$$z = \frac{\overline{x} - C}{s/\sqrt{n}}$$

$$t = \frac{\overline{x} - C}{s/\sqrt{n}}$$

TABLE I *(continued)*

Null Hypothesis $\mu_1 = \mu_2$

$$z = \frac{\bar{x}_1 - \bar{x}_2}{\sqrt{s_1^2/n_1 + s_2^2/n_2}}$$

$$t = \frac{\bar{x}_1 - \bar{x}_2}{\sqrt{s^2/n_1 + s^2/n_2}} \qquad s^2 = \frac{(n_1 - 1)s_1^2 + (n_2 - 1)s_2^2}{n_1 + n_2 - 2}$$

Null Hypothesis $\sigma_1 = \sigma_2$

$$F = \frac{s_1^2}{s_2^2}$$

Null Hypothesis sample is a random sample from a population with a given distribution.

$$\chi^2 = \frac{(f - F)^2}{F}$$

$$\text{df} = c - 1 \text{ or } (c-1)(r-1)$$

Null Hypothesis $\mu_1 = \mu_2 = \mu_3 = \text{etc.}$

ONE-WAY ANALYSIS OF VARIANCE

$$F = \frac{\text{between groups variance}}{\text{within groups variance}}$$

Total sum of squares

$$\Sigma x_T^2 - \frac{(\Sigma x_T)^2}{N}$$

$$\text{df} = N - 1$$

Between groups sum of squares

$$\frac{(\Sigma x_1)^2}{n_1} + \frac{(\Sigma x_2)^2}{n_2} + \ldots \text{etc.} - \frac{(\Sigma x_T)^2}{N}$$

$$\text{df} = g - 1$$

Within groups sum of squares
Total sum of squares − between groups sum of squares

or

$$\left[\Sigma x_1^2 - \frac{(\Sigma x_1)^2}{n_1}\right] + \left[\Sigma x_2^2 - \frac{(\Sigma x_2)^2}{n_2}\right] + \ldots \text{etc.}$$

$$\text{df} = N - g$$

TABLE I *(continued)*

TWO-WAY ANALYSIS OF VARIANCE

$$F = \frac{\text{between rows variance}}{\text{within groups variance}}$$

$$F = \frac{\text{between columns variance}}{\text{within groups variance}}$$

$$F = \frac{\text{interaction variance}}{\text{within groups variance}}$$

Total sum of squares

$$\Sigma x_T^2 - \frac{(\Sigma x_T)^2}{N}$$

Between groups sum of squares

$$\frac{(\Sigma x_1)^2}{n} + \frac{(\Sigma x_2)^2}{n} + \dots \text{etc.} - \frac{(\Sigma x_T)^2}{N}$$

$$df = rc - 1$$

Between rows sum of squares

$$\frac{(\Sigma x_{row1})^2}{nc} + \frac{(\Sigma x_{row2})^2}{nc} + \dots \text{etc.} - \frac{(\Sigma x_T)^2}{N}$$

$$df = r - 1$$

Between columns sum of squares

$$\frac{(\Sigma x_{col1})^2}{nr} + \frac{(\Sigma x_{col2})^2}{nr} + \dots \text{etc.} - \frac{(\Sigma x_T)^2}{N}$$

$$df = c - 1$$

Interaction sum of squares
Between groups sum of squares − [between rows sum of squares + between columns sum of squares]

or

$$\frac{(\Sigma x_1)^2}{n} + \frac{(\Sigma x_2)^2}{n} + \dots \text{etc.} - \frac{(\Sigma x_{row1})^2}{nc} - \frac{(\Sigma x_{row2})^2}{nc} - \dots \text{etc.} - \frac{(\Sigma x_{col1})^2}{nr}$$
$$- \frac{(\Sigma x_{col2})^2}{nr} - \dots \text{etc.} + \frac{(\Sigma x_T)^2}{N}$$

$$df = (r - 1)(c - 1)$$

Within groups sum of squares
Total sum of squares − between groups sum of squares

or

$$\left[\Sigma x_1^2 - \frac{(\Sigma x_1)^2}{n} \right] + \left[\Sigma x_2^2 - \frac{(\Sigma x_2)^2}{n} \right] + \dots \text{etc.}$$

$$df = N - rc$$

HOW TO USE TABLE II

To find the square root of a number between 1 and 10, find your number in the column headed **N**. Look across to the column headed $\sqrt{N}$ for the square root. For example, to find the square root of 5:

    locate 5.00 under **N**

    look across to find $\sqrt{N} = 2.23607$

To find the square root of a number from 10 to 100, divide the number by 10. Find the result in the column headed **N**. Then look across to the column headed $\sqrt{10N}$ for your square root. For example, to find the square root of 50:

    $50 \div 10 = 5$

    locate 5.00 in the **N** column

    look across to find $\sqrt{10N} = 7.07107$

To find the square root of a number over 100, move the decimal point an even number of places to the left until you have a number between 1 and 100. Look up the square root of that number in the table, as shown above. Then move the decimal point in the answer back to the right *half* as many places as you moved to the left. For example, to find the square root of 5000:

    move the decimal point *two* places to the left 50.00

    $\sqrt{50} = 7.07107$

    move the decimal point in the answer *one* place to the right

    70.7107

    $\sqrt{5000} = 70.7107$

To find the square root of a number less than one, move the decimal point an even number of places to the right until you have a number between 1 and 100. Look up the square root of that number in the table, as shown above. Then move the decimal point in the answer back to the left *half* as many places as you moved to the right. For example, to find the square root of 0.0710:

    move the decimal point *two* places to the right 07.10

    $\sqrt{7.10} = 2.66458$

    move the decimal point *one* place to the left .266458

    $\sqrt{0.0710} = 0.266458$

TABLE II   SQUARES AND SQUARE ROOTS

N	N^2	$\sqrt{N}$	$\sqrt{10N}$	N	N^2	$\sqrt{N}$	$\sqrt{10N}$
1.00	1.0000	1.00000	3.16228	**1.50**	2.2500	1.22474	3.87298
1.01	1.0201	1.00499	3.17805	1.51	2.2801	1.22882	3.88587
1.02	1.0404	1.00995	3.19374	1.52	2.3104	1.23288	3.89872
1.03	1.0609	1.01489	3.20936	1.53	2.3409	1.23693	3.91152
1.04	1.0816	1.01980	3.22490	1.54	2.3716	1.24097	3.92428
1.05	1.1025	1.02470	3.24037	1.55	2.4025	1.24499	3.93700
1.06	1.1236	1.02956	3.25576	1.56	2.4336	1.24900	3.94968
1.07	1.1449	1.03441	3.27109	1.57	2.4649	1.25300	3.96232
1.08	1.1664	1.03923	3.28634	1.58	2.4964	1.25698	3.97492
1.09	1.1881	1.04403	3.30151	1.59	2.5281	1.26095	3.98748
1.10	1.2100	1.04881	3.31662	**1.60**	2.5600	1.26491	4.00000
1.11	1.2321	1.05357	3.33167	1.61	2.5921	1.26886	4.01248
1.12	1.2544	1.05830	3.34664	1.62	2.6244	1.27279	4.02492
1.13	1.2769	1.06301	3.36155	1.63	2.6569	1.27671	4.03733
1.14	1.2996	1.06771	3.37639	1.64	2.6896	1.28062	4.04969
1.15	1.3225	1.07238	3.39116	1.65	2.7225	1.28452	4.06202
1.16	1.3456	1.07703	3.40588	1.66	2.7556	1.28841	4.07431
1.17	1.3689	1.08167	3.42053	1.67	2.7889	1.29228	4.08656
1.18	1.3924	1.08628	3.43511	1.68	2.8224	1.29615	4.09878
1.19	1.4161	1.09087	3.44964	1.69	2.8561	1.30000	4.11096
1.20	1.4400	1.09545	3.46410	**1.70**	2.8900	1.30384	4.12311
1.21	1.4641	1.10000	3.47851	1.71	2.9241	1.30767	4.13521
1.22	1.4884	1.10454	3.49285	1.72	2.9584	1.31149	4.14729
1.23	1.5129	1.10905	3.50714	1.73	2.9929	1.31529	4.15933
1.24	1.5376	1.11355	3.52136	1.74	3.0276	1.31909	4.17133
1.25	1.5625	1.11803	3.53553	1.75	3.0625	1.32288	4.18330
1.26	1.5876	1.12250	3.54965	1.76	3.0976	1.32665	4.19524
1.27	1.6129	1.12694	3.56371	1.77	3.1329	1.33041	4.20714
1.28	1.6384	1.13137	3.57771	1.78	3.1684	1.33417	4.21900
1.29	1.6641	1.13578	3.59166	1.79	3.2041	1.33791	4.23084
1.30	1.6900	1.14018	3.60555	**1.80**	3.2400	1.34164	4.24264
1.31	1.7161	1.14455	3.61939	1.81	3.2761	1.34536	4.25441
1.32	1.7424	1.14891	3.63318	1.82	3.3124	1.34907	4.26615
1.33	1.7689	1.15326	3.64692	1.83	3.3489	1.35277	4.27785
1.34	1.7956	1.15758	3.66060	1.84	3.3856	1.35647	4.28952
1.35	1.8225	1.16190	3.67423	1.85	3.4225	1.36015	4.30116
1.36	1.8496	1.16619	3.68782	1.86	3.4596	1.36382	4.31277
1.37	1.8769	1.17047	3.70135	1.87	3.4969	1.36748	4.32435
1.38	1.9044	1.17473	3.71484	1.88	3.5344	1.37113	4.33590
1.39	1.9321	1.17898	3.72827	1.89	3.5721	1.37477	4.34741
1.40	1.9600	1.18322	3.74166	**1.90**	3.6100	1.37840	4.35890
1.41	1.9881	1.18743	3.75500	1.91	3.6481	1.38203	4.37035
1.42	2.0164	1.19164	3.76829	1.92	3.6864	1.38564	4.38178
1.43	2.0449	1.19583	3.78153	1.93	3.7249	1.38924	4.39318
1.44	2.0736	1.20000	3.79473	1.94	3.7636	1.39284	4.40454
1.45	2.1025	1.20416	3.80789	1.95	3.8025	1.39642	4.41588
1.46	2.1316	1.20830	3.82099	1.96	3.8416	1.40000	4.42719
1.47	2.1609	1.21244	3.83406	1.97	3.8809	1.40357	4.43847
1.48	2.1904	1.21655	3.84708	1.98	3.9204	1.40712	4.44972
1.49	2.2201	1.22066	3.86005	1.99	3.9601	1.41067	4.46094
1.50	2.2500	1.22474	3.87298	**2.00**	4.0000	1.41421	4.47214
N	N^2	$\sqrt{N}$	$\sqrt{10N}$	N	N^2	$\sqrt{N}$	$\sqrt{10N}$

TABLE II *(continued)*

N	N²	√N	√10N		N	N²	√N	√10N
2.00	4.0000	1.41421	4.47214		**2.50**	6.2500	1.58114	5.00000
2.01	4.0401	1.41774	4.48330		2.51	6.3001	1.58430	5.00999
2.02	4.0804	1.42127	4.49444		2.52	6.3504	1.58745	5.01996
2.03	4.1209	1.42478	4.50555		2.53	6.4009	1.59060	5.02991
2.04	4.1616	1.42829	4.51664		2.54	6.4516	1.59374	5.03984
2.05	4.2025	1.43178	4.52769		2.55	6.5025	1.59687	5.04975
2.06	4.2436	1.43527	4.53872		2.56	6.5536	1.60000	5.05964
2.07	4.2849	1.43875	4.54973		2.57	6.6049	1.60312	5.06952
2.08	4.3264	1.44222	4.56070		2.58	6.6564	1.60624	5.07937
2.09	4.3681	1.44568	4.57165		2.59	6.7081	1.60935	5.08920
2.10	4.4100	1.44914	4.58258		**2.60**	6.7600	1.61245	5.09902
2.11	4.4521	1.45258	4.59347		2.61	6.8121	1.61555	5.10882
2.12	4.4944	1.45602	4.60435		2.62	6.8644	1.61864	5.11859
2.13	4.5369	1.45945	4.61519		2.63	6.9169	1.62173	5.12835
2.14	4.5796	1.46287	4.62601		2.64	6.9696	1.62481	5.13809
2.15	4.6225	1.46629	4.63681		2.65	7.0225	1.62788	5.14782
2.16	4.6656	1.46969	4.64758		2.66	7.0756	1.63095	5.15752
2.17	4.7089	1.47309	4.65833		2.67	7.1289	1.63401	5.16720
2.18	4.7524	1.47648	4.66905		2.68	7.1824	1.63707	5.17687
2.19	4.7961	1.47986	4.67974		2.69	7.2361	1.64012	5.18652
2.20	4.8400	1.48324	4.69042		**2.70**	7.2900	1.64317	5.19615
2.21	4.8841	1.48661	4.70106		2.71	7.3441	1.64621	5.20577
2.22	5.9284	1.48997	4.71169		2.72	7.3984	1.64924	5.21536
2.23	4.9729	1.49332	4.72229		2.73	7.4529	1.65227	5.22494
2.24	5.0176	1.49666	4.73286		2.74	7.5076	1.65529	5.23450
2.25	5.0625	1.50000	4.74342		2.75	7.5625	1.65831	5.24404
2.26	5.1076	1.50333	4.75395		2.76	7.6176	1.66132	5.25357
2.27	5.1529	1.50665	4.76445		2.77	7.6729	1.66433	5.26308
2.28	5.1984	1.50997	4.77493		2.78	7.7284	1.66733	5.27257
2.29	5.2441	1.51327	4.78539		2.79	7.7841	1.67033	5.28205
2.30	5.2900	1.51658	4.79583		**2.80**	7.8400	1.67332	5.29150
2.31	5.3361	1.51987	4.80625		2.81	7.8961	1.67631	5.30094
2.32	5.3824	1.52315	4.81664		2.82	7.9524	1.67929	5.31037
2.33	5.4289	1.52643	4.82701		2.83	8.0089	1.68226	5.31977
2.34	5.4756	1.52971	4.83735		2.84	8.0656	1.68523	5.32917
2.35	5.5225	1.53297	4.84768		2.85	8.1225	1.68819	5.33854
2.36	5.5696	1.53623	4.85798		2.86	8.1796	1.69115	5.34790
2.37	5.6169	1.53948	4.86826		2.87	8.2369	1.69411	5.35724
2.38	5.6644	1.54272	4.87852		2.88	8.2944	1.69706	5.36656
2.39	5.7121	1.54596	4.88876		2.89	8.3521	1.70000	5.37587
2.40	5.7600	1.54919	4.89898		**2.90**	8.4100	1.70294	5.38516
2.41	5.8081	1.55252	4.90918		2.91	8.4681	1.70587	5.39444
2.42	5.8564	1.55563	4.91935		2.92	8.5264	1.70880	5.40370
2.43	5.9049	1.55885	4.92950		2.93	8.5849	1.71172	5.41295
2.44	5.9536	1.56205	4.93964		2.94	8.6436	1.71464	5.42218
2.45	6.0025	1.56525	4.94975		2.95	8.7025	1.71756	5.43139
2.46	6.0516	1.56844	4.95984		2.96	8.7616	1.72047	5.44059
2.47	6.1009	1.57162	4.96991		2.97	8.8209	1.72337	5.44977
2.48	6.1054	1.57480	4.97996		2.98	8.8804	1.72627	5.45894
2.49	6.2001	1.57797	4.98999		2.99	8.9401	1.72916	5.46809
2.50	6.2500	1.58114	5.00000		**3.00**	9.0000	1.73205	5.47723
N	N²	√N	√10N		N	N²	√N	√10N

TABLE II *(continued)*

N	N²	√N	√10N
3.00	9.0000	1.73205	5.47723
3.01	9.0601	1.73494	5.48635
3.02	9.1204	1.73781	5.49545
3.03	9.1809	1.74069	5.50454
3.04	9.2416	1.74356	5.51362
3.05	9.3025	1.74642	5.52268
3.06	9.3636	1.74929	5.53173
3.07	9.4249	1.75214	5.54076
3.08	9.4864	1.75499	5.54977
3.09	9.5481	1.75784	5.55878
3.10	9.6100	1.76068	5.56776
3.11	9.6721	1.76352	5.57674
3.12	9.7344	1.76635	5.58570
3.13	9.7969	1.76918	5.59464
3.14	9.8596	1.77200	5.60357
3.15	9.9225	1.77482	5.61249
3.16	9.9856	1.77764	5.62139
3.17	10.0489	1.78045	5.63028
3.18	10.1124	1.78326	5.63915
3.19	10.1761	1.78606	5.64801
3.20	10.2400	1.78885	5.65685
3.21	10.3041	1.79165	5.66569
3.22	10.3684	1.79444	5.67450
3.23	10.4329	1.79722	5.68331
3.24	10.4976	1.80000	5.69210
3.25	10.5625	1.80278	5.70088
3.26	10.6276	1.80555	5.70964
3.27	10.6929	1.80831	5.71839
3.28	10.7584	1.81108	5.72713
3.29	10.8241	1.81384	5.73585
3.30	10.8900	1.81659	5.74456
3.31	10.9561	1.81934	5.75326
3.32	10.0224	1.82209	5.76194
3.33	11.0889	1.82483	5.77062
3.34	11.1556	1.82757	5.77927
3.35	11.2225	1.83030	5.78792
3.36	11.2896	1.83303	5.79655
3.37	11.3569	1.83576	5.80517
3.38	11.4244	1.83848	5.81378
3.39	11.4921	1.84120	5.82237
3.40	11.5600	1.84391	5.83095
3.41	11.6281	1.84662	5.83952
3.42	11.6964	1.84932	5.84808
3.43	11.7649	1.85203	5.85662
3.44	11.8336	1.85472	5.86515
3.45	11.9025	1.85742	5.87367
3.46	11.9716	1.86011	5.88218
3.47	12.0409	1.86279	5.89067
3.48	12.1104	1.86548	5.89915
3.49	12.1801	1.86815	5.90762
3.50	12.2500	1.87083	5.91608

N	N²	√N	√10N
3.50	12.2500	1.87083	5.91608
3.51	12.3201	1.87350	5.92453
3.52	12.3904	1.87617	5.93296
3.53	12.4609	1.87883	5.94138
3.54	12.5316	1.88149	5.94979
3.55	12.6025	1.88414	5.95819
3.56	12.6736	1.88680	5.96657
3.57	12.7449	1.88944	5.97495
3.58	12.8164	1.89209	5.98331
3.59	12.8881	1.89473	5.99166
3.60	12.9600	1.89737	6.00000
3.61	13.0321	1.90000	6.00833
3.62	13.1044	1.90263	6.01664
3.63	13.1769	1.90526	6.02495
3.64	13.2496	1.90788	6.03324
3.65	13.3225	1.91050	6.04152
3.66	13.3956	1.91311	6.04949
3.67	13.4689	1.91572	6.05805
3.68	13.5424	1.91833	6.06630
3.69	13.6161	1.92094	6.07454
3.70	13.6900	1.92354	6.08276
3.71	13.7641	1.92614	6.09098
3.72	13.8384	1.92873	6.09918
3.73	13.9129	1.93132	6.10737
3.74	13.9876	1.93391	6.11555
3.75	14.0625	1.93649	6.12372
3.76	14.1376	1.93907	6.13188
3.77	14.2129	1.94165	6.14003
3.78	14.2884	1.94422	6.14817
3.79	14.3641	1.94679	6.15630
3.80	14.4400	1.94936	6.16441
3.81	14.5161	1.95192	6.17252
3.82	14.5924	1.95448	6.18061
3.83	14.6689	1.95704	6.18870
3.84	14.7456	1.95959	6.19677
3.85	14.8225	1.96214	6.20484
3.86	14.8996	1.96469	6.21289
3.87	14.9769	1.96723	6.22093
3.88	15.0544	1.96977	6.22896
3.89	15.1321	1.97231	6.23699
3.90	15.2100	1.97484	6.24500
3.91	15.2881	1.97737	6.25300
3.92	15.3664	1.97990	6.26099
3.93	15.4449	1.98242	6.26897
3.94	15.5236	1.98494	6.27694
3.95	15.6025	1.98746	6.28490
3.96	15.6816	1.98997	6.29285
3.97	15.7609	1.99249	6.30079
3.98	15.8404	1.99499	6.30872
3.99	15.9201	1.99750	6.31644
4.00	16.0000	2.00000	6.32456

N	N²	√N	√10N

TABLE II *(continued)*

N	N²	√N	√10N		N	N²	√N	√10N
4.00	16.0000	2.00000	6.32456		**4.50**	20.2500	2.12132	6.70820
4.01	16.0801	2.00250	6.33246		4.51	20.3401	2.12368	6.71565
4.02	16.1604	2.00499	6.34035		4.52	20.4304	2.12603	6.72309
4.03	16.2409	2.00749	6.34823		4.53	20.5209	2.12838	6.73053
4.04	16.3216	2.00998	6.35610		4.54	20.6116	2.13073	6.73795
4.05	16.4025	2.01246	6.36396		4.55	20.7025	2.13307	6.74537
4.06	16.4836	2.01494	6.37181		4.56	20.7936	2.13542	6.75278
4.07	16.5649	2.01742	6.37966		4.57	20.8849	2.13776	6.76018
4.08	16.6464	2.01990	6.38749		4.58	20.9764	2.14009	6.76757
4.09	16.7281	2.02237	6.39531		4.59	21.0681	2.14243	6.77495
4.10	16.8100	2.02485	6.40312		**4.60**	21.1600	2.14476	6.78233
4.11	16.8921	2.02731	6.41093		4.61	21.2521	2.14709	6.78970
4.12	16.9744	2.02978	6.41872		4.62	21.3444	2.14942	6.79706
4.13	17.0569	2.03224	6.42651		4.63	21.4369	2.15174	6.80441
4.14	17.1396	2.03470	6.43428		4.64	21.5296	2.15407	6.81175
4.15	17.2225	2.03715	6.44205		4.65	21.6225	2.15639	6.81909
4.16	17.3056	2.03961	6.44981		4.66	21.7156	2.15870	6.82642
4.17	17.3889	2.04206	6.45755		4.67	21.8089	2.16102	6.83374
4.18	17.4724	2.04450	6.46529		4.68	21.9024	2.16333	6.84105
4.19	17.5561	2.04695	6.47302		4.69	21.9961	2.16564	6.84836
4.20	17.6400	2.04939	6.48074		**4.70**	22.0900	2.16795	6.85565
4.21	17.7241	2.05183	6.48845		4.71	22.1841	2.17025	6.86294
4.22	17.8084	2.05426	6.49615		4.72	22.2784	2.17256	6.87023
4.23	17.8929	2.05670	6.50384		4.73	22.3729	2.17486	6.87750
4.24	17.9776	2.05913	6.51153		4.74	22.4676	2.17715	6.88477
4.25	18.0625	2.06155	6.51920		4.75	22.5625	2.17945	6.89202
4.26	18.1476	2.06398	6.52687		4.76	22.6576	2.18174	6.89928
4.27	18.2329	2.06640	6.53452		4.77	22.7529	2.18403	6.90652
4.28	18.3184	2.06882	6.54217		4.78	22.8484	2.18632	6.91375
4.29	18.4041	2.07123	6.54981		4.79	22.9441	2.18861	6.92098
4.30	18.4900	2.07364	6.55744		**4.80**	23.0400	2.19089	6.92820
4.31	18.5761	2.07605	6.66506		4.81	23.1361	2.19317	6.93542
4.32	18.6624	2.07846	6.57267		4.82	23.2324	2.19545	6.94262
4.33	18.7489	2.08087	6.58027		4.83	23.3289	2.19773	6.94982
4.34	18.8356	2.08327	6.58787		4.84	23.4256	2.20000	6.95701
4.35	18.9225	2.08567	6.59545		4.85	23.5225	2.20227	6.96419
4.36	19.0096	2.08806	6.60303		4.86	23.6196	2.20454	6.97137
4.37	19.0969	2.09045	6.61060		4.87	23.7169	2.20681	6.97854
4.38	19.1844	2.09284	6.61816		4.88	23.8144	2.20907	6.98570
4.39	19.2721	2.09523	6.62571		4.89	23.9121	2.21133	6.99285
4.40	19.3600	2.09762	6.63325		**4.90**	24.0100	2.21359	7.00000
4.41	19.4481	2.10000	6.64078		4.91	24.1081	2.21585	7.00714
4.42	19.5364	2.10238	6.64831		4.92	24.2064	2.21811	7.01427
4.43	19.6249	2.10476	6.65582		4.93	24.3049	2.22036	7.02140
4.44	19.7136	2.10713	6.66333		4.94	24.4036	2.22261	7.02851
4.45	19.8025	2.10950	6.67083		4.95	24.5025	2.22486	7.03562
4.46	19.8916	2.11187	6.67832		4.96	24.6016	2.22711	7.04273
4.47	19.9809	2.11424	6.68581		4.97	24.7009	2.22935	7.04982
4.48	20.0704	2.11660	6.69328		4.98	24.8004	2.23159	7.05691
4.49	20.1601	2.11896	6.70075		4.99	24.9001	2.23383	7.06399
4.50	20.2500	2.12132	6.70820		**5.00**	25.0000	2.23607	7.07107
N	N²	√N	√10N		N	N²	√N	√10N

TABLE II *(continued)*

N	N²	$\sqrt{N}$	$\sqrt{10N}$		N	N²	$\sqrt{N}$	$\sqrt{10N}$
5.00	25.0000	2.23607	7.07107		**5.50**	30.2500	2.34521	7.41620
5.01	25.1001	2.23830	7.07814		5.51	30.3601	2.34734	7.42294
5.02	25.2004	2.24054	7.08520		5.52	30.4704	2.34947	7.42967
5.03	25.3009	2.24277	7.09225		5.53	30.5809	2.35160	7.43640
5.04	25.4016	2.24499	7.09930		5.54	30.6916	2.35372	7.44312
5.05	25.5025	2.24722	7.10634		5.55	30.8025	2.35584	7.44983
5.06	25.6036	2.24944	7.11337		5.56	30.9136	2.35797	7.45654
5.07	25.7049	2.25167	7.12039		5.57	31.0249	2.36008	7.46324
5.08	25.8064	2.25389	7.12741		5.58	31.1364	2.36220	7.46994
5.09	25.9081	2.25610	7.13442		5.59	31.2481	2.36432	7.47663
5.10	26.0100	2.25832	7.14143		**5.60**	31.3600	2.36643	7.48331
5.11	26.1121	2.26053	7.14843		5.61	31.4721	2.36854	7.48999
5.12	26.2144	2.26274	7.15542		5.62	31.5844	2.37065	7.49667
5.13	26.3169	2.26495	7.16240		5.63	31.6969	2.37276	7.50333
5.14	26.4196	2.26716	7.16938		5.64	31.8096	2.37487	7.50999
5.15	26.5225	2.26936	7.17635		5.65	31.9225	2.37697	7.51665
5.16	26.6256	2.27156	7.18331		5.66	32.0356	2.37908	7.52330
5.17	26.7289	2.27376	7.19027		5.67	32.1489	2.38118	7.52994
5.18	26.8324	2.27596	7.19722		5.68	32.2624	2.38328	7.53658
5.19	26.9361	2.27816	7.20417		5.69	32.3761	2.38537	7.54321
5.20	27.0400	2.28035	7.21110		**5.70**	32.4900	2.38747	7.54983
5.21	27.1441	2.28254	7.21803		5.71	32.6041	2.38956	7.55645
5.22	27.2484	2.28473	7.22496		5.72	32.7184	2.39165	7.56307
5.23	27.3529	2.28692	7.23187		5.73	32.8329	2.39374	7.56968
5.24	27.4576	2.28910	7.23838		5.74	32.9476	2.39583	7.57628
5.25	27.5625	2.29129	7.24569		5.75	33.0625	2.39792	7.58288
5.26	27.6676	2.29347	7.25259		5.76	33.1776	2.40000	7.58947
5.27	27.7729	2.29565	7.25948		5.77	33.2929	2.40208	7.59605
5.28	27.8784	2.29783	7.26636		5.78	33.4084	2.40416	7.60263
5.29	27.9841	2.30000	7.27324		5.79	33.5241	2.40624	7.60920
5.30	28.0900	2.30217	7.28011		**5.80**	33.6400	2.40832	7.61577
5.31	28.1961	2.30434	7.28697		5.81	33.7561	2.41039	7.62234
5.32	28.3024	2.30651	7.29383		5.82	33.8724	2.41247	7.62889
5.33	28.4089	2.30868	7.30068		5.83	33.9889	2.41454	7.63544
5.34	28.5156	2.31084	7.30753		5.84	34.1056	2.41661	7.64199
5.35	28.6225	2.31301	7.31437		5.85	34.2225	2.41868	7.64853
5.36	28.7296	2.31517	7.32120		5.86	34.3396	2.42074	7.65506
5.37	28.8369	2.31733	7.32803		5.87	34.4569	2.42281	7.66159
5.38	28.9444	2.31948	7.33485		5.88	34.5744	2.42487	7.66812
5.39	29.0521	2.32164	7.34166		5.89	34.6921	2.42693	7.67463
5.40	29.1600	2.32379	7.34847		**5.90**	34.8100	2.42899	7.68115
5.41	29.2681	2.32594	7.35527		5.91	34.9281	2.43105	7.68765
5.42	29.3764	2.32809	7.36206		5.92	35.0464	2.43311	7.69415
5.43	29.4849	2.33024	7.36885		5.93	35.1649	2.43516	7.70065
5.44	29.5936	2.33238	7.37564		5.94	35.2836	2.43721	7.70714
5.45	29.7025	2.33452	7.38241		5.95	35.4025	2.43926	7.71362
5.46	29.8116	2.33666	7.38918		5.96	35.5216	2.44131	7.72010
5.47	29.9209	2.33880	7.39594		5.97	35.6409	2.44336	7.72658
5.48	30.0304	2.34094	7.40270		5.98	35.7604	2.44540	7.73305
5.49	30.1401	2.34307	7.40945		5.99	35.8801	2.44745	7.73951
5.50	30.2500	2.34521	7.41620		**6.00**	36.0000	2.44949	7.74597
N	N²	$\sqrt{N}$	$\sqrt{10N}$		N	N²	$\sqrt{N}$	$\sqrt{10N}$

TABLE II *(continued)*

N	N²	√N	√10N		N	N²	√N	√10N
6.00	36.0000	2.44949	7.74597		**6.50**	42.2500	2.54951	8.06226
6.01	36.1201	2.45153	7.75242		6.51	42.3801	2.55147	8.06846
6.02	36.2404	2.45357	7.75887		6.52	42.5104	2.55343	8.07465
6.03	36.3609	2.45561	7.76531		6.53	42.6409	2.55539	8.08084
6.04	36.4816	2.45764	7.77174		6.54	42.7716	2.55734	8.08703
6.05	36.6025	2.45967	7.77817		6.55	42.9025	2.55930	8.09321
6.06	36.7236	2.46171	7.78460		6.56	43.0336	2.56125	8.09938
6.07	36.8449	2.46374	7.79102		6.57	43.1649	2.56320	8.10555
6.08	36.9664	2.46577	7.79744		6.58	43.2964	2.56515	8.11172
6.09	37.0881	2.46779	7.80385		6.59	43.4281	2.56710	8.11788
6.10	37.2100	2.46982	7.81025		**6.60**	43.5600	2.56905	8.12404
6.11	37.3321	2.47184	7.81665		6.61	43.6921	2.57099	8.13019
6.12	37.4544	2.47386	7.82304		6.62	43.8244	2.57294	8.13634
6.13	37.5769	2.47588	7.82943		6.63	43.9569	2.57488	8.14248
6.14	37.6996	2.47790	7.83582		6.64	44.0896	2.57682	8.14862
6.15	37.8225	2.47992	7.84219		6.65	44.2225	2.57876	8.15475
6.16	37.9456	2.48193	7.84857		6.66	44.3556	2.58070	8.16088
6.17	38.0689	2.48395	7.85493		6.67	44.4889	2.58263	8.16701
6.18	38.1924	2.48596	7.86130		6.68	44.6224	2.58457	8.17313
6.19	38.3161	2.48797	7.86766		6.69	44.7561	2.58650	8.17924
6.20	38.4400	2.48998	7.87401		**6.70**	44.8900	2.58844	8.18535
6.21	38.5641	2.49199	7.88036		6.71	45.0241	2.59037	8.19146
6.22	38.6884	2.49399	7.88670		6.72	45.1584	2.59230	8.19756
6.23	38.8129	2.49600	7.89303		6.73	45.2929	2.59422	8.20366
6.24	38.9376	2.49800	7.89937		6.74	45.4276	2.59615	8.20975
6.25	39.0625	2.50000	7.90569		6.75	45.5625	2.59808	8.21584
6.26	39.1876	2.50200	7.91202		6.76	45.6976	2.60000	8.22192
6.27	39.3129	2.50400	7.91833		6.77	45.8329	2.60192	8.22800
6.28	39.4384	2.50599	7.92465		6.78	45.9684	2.60384	8.23408
6.29	39.5641	2.50799	7.93095		6.79	46.1041	2.60576	8.24015
6.30	39.6900	2.50998	7.93725		**6.80**	46.2400	2.60768	8.24621
6.31	39.8161	2.51197	7.94355		6.81	46.3761	2.60960	8.25227
6.32	39.9424	2.51396	7.94984		6.82	46.5124	2.61151	8.25833
6.33	40.0689	2.51595	7.95613		6.83	46.6489	2.61343	8.26438
6.34	40.1956	2.51794	7.96241		6.84	46.7856	2.61534	8.27043
6.35	40.3225	2.51992	7.96869		6.85	46.9225	2.61725	8.27647
6.36	40.4496	2.52190	7.97496		6.86	47.0596	2.61916	8.28251
6.37	40.5769	2.52389	7.98123		6.87	47.1969	2.62107	8.28855
6.38	40.7044	2.52587	7.98749		6.88	47.3344	2.62298	8.29458
6.39	40.8321	2.52784	7.99375		6.89	47.4721	2.62488	8.30060
6.40	40.9600	2.52982	8.00000		**6.90**	47.6100	2.62679	8.30662
6.41	41.0881	2.53180	8.00625		6.91	47.7481	2.62869	8.31264
6.42	41.2164	2.53377	8.01249		6.92	47.8864	2.63059	8.31865
6.43	41.3449	2.53574	8.01873		6.93	48.0249	2.63249	8.32466
6.44	41.4736	2.53772	8.02496		6.94	48.1636	2.63439	8.33067
6.45	41.6025	2.53969	8.03119		6.95	48.3025	2.63629	8.33667
6.46	41.7316	2.54165	8.03741		6.96	48.4416	2.63818	8.34266
6.47	41.8609	2.54362	8.04363		6.97	48.5809	2.64008	8.34865
6.48	41.9904	2.54558	8.04984		6.98	48.7204	2.64197	8.35464
6.49	42.1201	2.54755	8.05605		6.99	48.8601	2.64386	8.36062
6.50	42.2500	2.54951	8.06226		**7.00**	49.0000	2.64575	8.36660
N	N²	√N	√10N		N	N²	√N	√10N

TABLE II *(continued)*

N	N²	√N	√10N
7.00	49.0000	2.64575	8.36660
7.01	49.1401	2.64764	8.37257
7.02	49.2804	2.64953	8.37854
7.03	49.4209	2.65141	8.38451
7.04	49.5616	2.65330	8.39047
7.05	49.7025	2.65518	8.39643
7.06	49.8436	2.65707	8.40238
7.07	49.9849	2.65895	8.40833
7.08	50.1264	2.66083	8.41427
7.09	50.2681	2.66271	8.42021
7.10	50.4100	2.66458	8.42615
7.11	50.5521	2.66646	8.43208
7.12	50.6944	2.66833	8.43801
7.13	50.8369	2.67021	8.44393
7.14	50.9796	2.67208	8.44985
7.15	51.1225	2.67395	8.45577
7.16	51.2656	2.67582	8.46168
7.17	51.4089	2.67769	8.46759
7.18	51.5524	2.67955	8.47349
7.19	51.6961	2.68142	8.47939
7.20	51.8400	2.68328	8.48528
7.21	51.9841	2.68514	8.49117
7.22	52.1284	2.68701	8.49706
7.23	52.2729	2.68887	8.50294
7.24	52.4176	2.69072	8.50882
7.25	52.5625	2.69258	8.51469
7.26	52.7076	2.69444	8.52056
7.27	52.8529	2.69629	8.52643
7.28	52.9984	2.69815	8.53229
7.29	53.1441	2.70000	8.53815
7.30	53.2900	2.70185	8.54400
7.31	53.4361	2.70370	8.54985
7.32	53.5824	2.70555	8.55570
7.33	53.7289	2.70740	8.56154
7.34	53.8756	2.70924	8.56738
7.35	54.0225	2.71109	8.57321
7.36	54.1696	2.71293	8.57904
7.37	54.3169	2.71477	8.58487
7.38	54.4644	2.71662	8.59069
7.39	54.6121	2.71846	8.59651
7.40	54.7600	2.72029	8.60233
7.41	54.9081	2.72213	8.60814
7.42	55.0564	2.72397	8.61394
7.43	55.2049	2.72580	8.61974
7.44	55.3536	2.72764	8.62554
7.45	55.5025	2.72947	8.63134
7.46	55.6516	2.73130	8.63713
7.47	55.8009	2.73313	8.64292
7.48	55.9504	2.73496	8.64870
7.49	56.1001	2.73679	8.65448
7.50	56.2500	2.73861	8.66025

N	N²	√N	√10N
7.50	56.2500	2.73861	8.66025
7.51	56.4001	2.74044	8.66603
7.52	56.5504	2.74226	8.67179
7.53	56.7009	2.74408	8.67756
7.54	56.8516	2.74591	8.68332
7.55	57.0025	2.74773	8.68907
7.56	57.1536	2.74955	8.69483
7.57	57.3049	2.75136	8.70057
7.58	57.4564	2.75318	8.70632
7.59	57.6081	2.75500	8.71206
7.60	57.7600	2.75681	8.71780
7.61	57.9121	2.75862	8.72353
7.62	58.0644	2.76043	8.72926
7.63	58.2169	2.76225	8.73499
7.64	58.3696	2.76405	8.74071
7.65	58.5225	2.76586	8.74643
7.66	58.6756	2.76767	8.75214
7.67	58.8289	2.76948	8.75785
7.68	58.9824	2.77128	8.76356
7.69	59.1361	2.77308	8.76926
7.70	59.2900	2.77489	8.77496
7.71	59.4441	2.77669	8.78066
7.72	59.5984	2.77849	8.78635
7.73	59.7529	2.78029	8.79204
7.74	59.9076	2.78209	8.79773
7.75	60.0625	2.78388	8.80341
7.76	60.2176	2.78568	8.80909
7.77	60.3729	2.78747	8.81476
7.78	60.5284	2.78927	8.82043
7.79	60.6841	2.79106	8.82610
7.80	60.8400	2.79285	8.83176
7.81	60.9961	2.79464	8.83742
7.82	61.1524	2.79643	8.84308
7.83	61.3089	2.79821	8.84873
7.84	61.4656	2.80000	8.85438
7.85	61.6225	2.80179	8.86002
7.86	61.7796	2.80357	8.86566
7.87	61.9369	2.80535	8.87130
7.88	62.0944	2.80713	8.87694
7.89	62.2521	2.80891	8.88257
7.90	62.4100	2.81069	8.88819
7.91	62.5681	2.81247	8.89382
7.92	62.7264	2.81425	8.89944
7.93	62.8849	2.81603	8.90505
7.94	63.0436	2.81780	8.91067
7.95	63.2025	2.81957	8.91628
7.96	63.3616	2.82135	8.92188
7.97	63.5209	2.82312	8.92749
7.98	63.6804	2.82489	8.93308
7.99	63.8401	2.82666	8.93868
8.00	64.0000	2.82843	8.94427

N	N²	√N	√10N

TABLE II *(continued)*

N	N²	√N̄	√10N̄		N	N²	√N̄	√10N̄
8.00	64.0000	2.82843	8.94427		**8.50**	72.2500	2.91548	9.21954
8.01	64.1601	2.83019	8.94986		8.51	72.4201	2.91719	9.22497
8.02	64.3204	2.83196	8.95545		8.52	72.5904	2.91890	9.23038
8.03	64.4809	2.83373	8.96103		8.53	72.7609	2.92062	9.23580
8.04	64.6416	2.83549	8.96660		8.54	72.9316	2.92233	9.24121
8.05	64.8025	2.83725	8.97218		8.55	73.1025	2.92404	9.24662
8.06	64.9636	2.83901	8.97775		8.56	73.2736	2.92575	9.25203
8.07	65.1249	2.84077	8.98332		8.57	73.4449	2.92746	9.25743
8.08	65.2864	2.84253	8.98888		8.58	73.6164	2.92916	9.26283
8.09	65.4481	2.84429	8.99444		8.59	73.7881	2.93087	9.26823
8.10	65.6100	2.84605	9.00000		**8.60**	73.9600	2.93258	9.27362
8.11	65.7721	2.84781	9.00555		8.61	74.1321	2.93428	9.27901
8.12	65.9344	2.84956	9.01110		8.62	74.3044	2.93598	9.28440
8.13	66.0969	2.85132	9.01665		8.63	74.4769	2.93769	9.28978
8.14	66.2596	2.85307	9.02219		8.64	74.6496	2.93939	9.29516
8.15	66.4225	2.85482	9.02774		8.65	74.8225	2.94109	9.30054
8.16	66.5856	2.85657	9.03327		8.66	74.9956	2.94279	9.30591
8.17	66.7489	2.85832	9.03881		8.67	75.1689	2.94449	9.31128
8.18	66.9124	2.86007	9.04434		8.68	75.3424	2.94618	9.31665
8.19	67.0761	2.86182	9.04986		8.69	75.5161	2.94788	9.32202
8.20	67.2400	2.86356	9.05539		**8.70**	75.6900	2.94958	9.32738
8.21	67.4041	2.86531	9.06091		8.71	75.8641	2.95127	9.33274
8.22	67.5684	2.86705	9.06642		8.72	76.0384	2.95296	9.33809
8.23	67.7329	2.86880	9.07193		8.73	76.2129	2.95466	9.34345
8.24	67.8976	2.87054	9.07744		8.74	76.3876	2.95635	9.34880
8.25	68.0625	2.87228	9.08295		8.75	76.5625	2.95804	9.35414
8.26	68.2276	2.87402	9.08845		8.76	76.7376	2.95973	9.35949
8.27	68.3929	2.87576	9.09395		8.77	76.9129	2.96142	9.36483
8.28	68.5584	2.87750	9.09945		8.78	77.0884	2.96311	9.37017
8.29	68.7241	2.87924	9.10494		8.79	77.2641	2.96479	9.37550
8.30	68.8900	2.88097	9.11045		**8.80**	77.4400	2.96648	9.38083
8.31	69.0561	2.88271	9.11592		8.81	77.6161	2.96816	9.38616
8.32	69.2224	2.88444	9.12140		8.82	77.7924	2.96985	9.39149
8.33	69.3889	2.88617	9.12688		8.83	77.9689	2.97153	9.39681
8.34	69.5556	2.88791	9.13236		8.84	78.1456	2.97321	9.40213
8.35	69.7225	2.88964	9.13783		8.85	78.3225	2.97489	9.40744
8.36	69.8896	2.89137	9.14330		8.86	78.4996	2.97658	9.41276
8.37	70.0569	2.89310	9.14877		8.87	78.6769	2.97825	9.41807
8.38	70.2244	2.89482	9.15423		8.88	78.8544	2.97993	9.42338
8.39	70.3921	2.89655	9.15969		8.89	79.0321	2.98161	9.42868
8.40	70.5600	2.89828	9.16515		**8.90**	79.2100	2.98329	9.43398
8.41	70.7281	2.90000	9.17061		8.91	79.3881	2.98496	9.43928
8.42	70.8964	2.90172	9.17606		8.92	79.5664	2.98664	9.44458
8.43	71.0649	2.90345	9.18150		8.93	79.7449	2.98831	9.44987
8.44	71.2336	2.90517	9.18695		8.94	79.9236	2.98998	9.45516
8.45	71.4025	2.90689	9.19239		8.95	80.1025	2.99166	9.46044
8.46	71.5716	2.90861	9.19783		8.96	80.2816	2.99333	9.46573
8.47	71.7409	2.91033	9.20326		8.97	80.4609	2.99500	9.47101
8.48	71.9104	2.91204	9.20869		8.98	80.6404	2.99666	9.47629
8.49	72.0801	2.91376	9.21412		8.99	80.8201	2.99833	9.48156
8.50	72.2500	2.91548	9.21954		**9.00**	81.0000	3.00000	9.48683
N	N²	√N̄	√10N̄		N	N²	√N̄	√10N̄

TABLE II *(continued)*

N	N²	√N	√10N
9.00	81.0000	3.00000	9.48683
9.01	81.1801	3.00167	9.49210
9.02	81.3604	3.00333	9.49737
9.03	81.5409	3.00500	9.50263
9.04	81.7216	3.00666	9.50789
9.05	81.9025	3.00832	9.51315
9.06	82.0836	3.00998	9.51840
9.07	82.2649	3.01164	9.52365
9.08	82.4464	3.01330	9.52890
9.09	82.6281	3.01496	9.53415
9.10	82.8100	3.01662	9.53939
9.11	82.9921	3.01828	9.54463
9.12	83.1744	3.01993	9.54987
9.13	83.3569	3.02159	9.55510
9.14	83.5396	3.02324	9.56033
9.15	83.7225	3.02490	9.56556
9.16	83.9056	3.02655	9.57079
9.17	84.0889	3.02820	9.57601
9.18	84.2724	3.02985	9.58123
9.19	84.4561	3.03150	9.58645
9.20	84.6400	3.03315	9.59166
9.21	84.8241	3.03480	9.59687
9.22	85.0084	3.03645	9.60208
9.23	85.1929	3.03809	9.60729
9.24	85.3776	3.03974	9.61249
9.25	85.5625	3.04138	9.61769
9.26	85.7476	3.04302	9.62289
9.27	85.9329	3.04467	9.62808
9.28	86.1184	3.04631	9.63328
9.29	86.3041	3.04795	9.63846
9.30	86.4900	3.04959	9.64365
9.31	86.6761	3.05123	9.64883
9.32	86.8624	3.05287	9.65401
9.33	87.0489	3.05450	9.65919
9.34	87.2356	3.05614	9.66437
9.35	87.4225	3.05778	9.66954
9.36	87.6096	3.05941	9.67471
9.37	87.7969	3.06105	9.67988
9.38	87.9844	3.06268	9.68504
9.39	88.1721	3.06431	9.69020
9.40	88.3600	3.06594	9.69536
9.41	88.5481	3.06757	9.70052
9.42	88.7364	3.06920	9.70567
9.43	88.9249	3.07083	9.71082
9.44	89.1136	3.07246	9.71597
9.45	89.3025	3.07409	9.72111
9.46	89.4916	3.07571	9.72625
9.47	89.6809	3.07734	9.73139
9.48	89.8704	3.07896	9.73653
9.49	90.0601	3.08058	9.74166
9.50	90.2500	3.08221	9.74679

N	N²	√N	√10N
9.50	90.2500	3.08221	9.74679
9.51	90.4401	3.08383	9.75192
9.52	90.6304	3.08545	9.75705
9.53	90.8209	3.08707	9.76217
9.54	91.0116	3.08869	9.76729
9.55	91.2025	3.09031	9.77241
9.56	91.3936	3.09192	9.77753
9.57	91.5849	3.09354	9.78264
9.58	91.7764	3.09516	9.78775
9.59	91.9681	3.09677	9.79285
9.60	92.1600	3.09839	9.79796
9.61	92.3521	3.10000	9.80306
9.62	92.5444	3.10161	9.80816
9.63	92.7369	3.10322	9.81326
9.64	92.9296	3.10483	9.81835
9.65	93.1225	3.10644	9.82344
9.66	93.3156	3.10805	9.82853
9.67	93.5089	3.10966	9.83362
9.68	93.7024	3.11127	9.83870
9.69	93.8961	3.11288	9.84378
9.70	94.0900	3.11448	9.84886
9.71	94.2841	3.11609	9.85393
9.72	94.4784	3.11769	9.85901
9.73	94.6729	3.11929	9.86408
9.74	94.8676	3.12090	9.86914
9.75	95.0625	3.12250	9.87421
9.76	95.2576	3.12410	9.87927
9.77	95.4529	3.12570	9.88433
9.78	95.6484	3.12730	9.88939
9.79	95.8441	3.12890	9.89444
9.80	96.0400	3.13050	9.89949
9.81	96.2361	3.13209	9.90454
9.82	96.4324	3.13369	9.90959
9.83	96.6289	3.13528	9.91464
9.84	96.8256	3.13688	9.91968
9.85	97.0225	3.13847	9.92472
9.86	97.2196	3.14006	9.92974
9.87	97.4169	3.14166	9.93479
9.88	97.6144	3.14325	9.93982
9.89	97.8121	3.14484	9.94485
9.90	98.0100	3.14643	9.94987
9.91	98.2081	3.14802	9.95490
9.92	98.4064	3.14960	9.95992
9.93	98.6049	3.15119	9.96494
9.94	98.8036	3.15278	9.96995
9.95	99.0025	3.15436	9.97497
9.96	99.2016	3.15595	9.97998
9.97	99.4009	3.15753	9.98499
9.98	99.6004	3.15911	9.98999
9.99	99.8001	3.16070	9.99500
10.0	100.000	3.16228	10.0000

N	N²	√N	√10N

From Paul G. Hoel, *Elementary Statistics*, 3rd ed., © 1971, John Wiley and Sons, Inc., New York, pp. 275-283.

TABLE III   BINOMIAL PROBABILITIES

n	x	0.05	0.1	0.2	0.25	0.3	0.4	P 0.5	0.6	0.7	0.75	0.8	0.9	0.95
2	0	0.902	0.810	0.640	0.563	0.490	0.360	0.250	0.160	0.090	0.063	0.040	0.010	0.002
	1	0.095	0.180	0.320	0.375	0.420	0.480	0.500	0.480	0.420	0.375	0.320	0.180	0.095
	2	0.002	0.010	0.040	0.063	0.090	0.160	0.250	0.360	0.490	0.563	0.640	0.810	0.902
3	0	0.857	0.729	0.512	0.422	0.343	0.216	0.125	0.064	0.027	0.016	0.008	0.001	
	1	0.135	0.243	0.384	0.422	0.441	0.432	0.375	0.288	0.189	0.141	0.096	0.027	0.007
	2	0.007	0.027	0.096	0.141	0.189	0.288	0.375	0.432	0.441	0.422	0.384	0.243	0.135
	3		0.001	0.008	0.016	0.027	0.064	0.125	0.216	0.343	0.422	0.512	0.729	0.857
4	0	0.815	0.656	0.410	0.316	0.240	0.130	0.062	0.026	0.008	0.004	0.002		
	1	0.171	0.292	0.410	0.422	0.412	0.346	0.250	0.154	0.076	0.047	0.026	0.004	
	2	0.014	0.049	0.154	0.211	0.265	0.346	0.375	0.346	0.265	0.211	0.154	0.049	0.014
	3		0.004	0.026	0.047	0.076	0.154	0.250	0.346	0.412	0.422	0.410	0.292	0.171
	4			0.002	0.004	0.008	0.026	0.062	0.130	0.240	0.316	0.410	0.656	0.815
5	0	0.774	0.590	0.328	0.237	0.168	0.078	0.031	0.010	0.002	0.001			
	1	0.204	0.328	0.410	0.396	0.360	0.259	0.156	0.007	0.028	0.015	0.006		
	2	0.021	0.073	0.205	0.264	0.309	0.346	0.312	0.230	0.132	0.088	0.051	0.008	0.001
	3	0.001	0.008	0.051	0.088	0.132	0.230	0.312	0.346	0.309	0.274	0.205	0.073	0.021
	4			0.006	0.015	0.028	0.077	0.156	0.259	0.360	0.396	0.410	0.328	0.204
	5				0.001	0.002	0.010	0.031	0.078	0.168	0.237	0.328	0.590	0.774
6	0	0.735	0.531	0.262	0.178	0.118	0.047	0.016	0.004	0.001				
	1	0.232	0.354	0.393	0.356	0.303	0.187	0.094	0.037	0.010	0.004	0.002		
	2	0.031	0.098	0.246	0.297	0.324	0.311	0.234	0.138	0.060	0.033	0.015	0.001	
	3	0.002	0.015	0.082	0.132	0.185	0.276	0.132	0.276	0.185	0.132	0.082	0.015	0.002
	4		0.001	0.015	0.033	0.060	0.138	0.234	0.311	0.324	0.297	0.246	0.098	0.031
	5			0.002	0.004	0.010	0.037	0.094	0.187	0.303	0.356	0.393	0.354	0.232
	6					0.001	0.004	0.016	0.047	0.118	0.178	0.262	0.531	0.735
7	0	0.698	0.478	0.210	0.134	0.082	0.028	0.008	0.002					
	1	0.257	0.372	0.367	0.312	0.247	0.131	0.055	0.017	0.004	0.001			
	2	0.041	0.124	0.275	0.312	0.318	0.261	0.164	0.077	0.025	0.012	0.004		
	3	0.004	0.023	0.115	0.173	0.227	0.290	0.273	0.194	0.097	0.058	0.029	0.003	
	4		0.003	0.029	0.058	0.097	0.194	0.273	0.290	0.227	0.173	0.115	0.023	0.004
	5			0.004	0.012	0.025	0.077	0.164	0.261	0.318	0.312	0.275	0.124	0.041
	6				0.001	0.004	0.017	0.055	0.131	0.247	0.312	0.367	0.372	0.257
	7						0.002	0.008	0.028	0.082	0.134	0.210	0.478	0.698
8	0	0.663	0.430	0.168	0.100	0.058	0.017	0.004	0.001					
	1	0.279	0.383	0.336	0.267	0.198	0.090	0.031	0.008	0.001				
	2	0.051	0.149	0.294	0.312	0.296	0.209	0.109	0.041	0.010	0.004	0.001		
	3	0.005	0.033	0.147	0.208	0.254	0.279	0.219	0.124	0.047	0.023	0.009		
	4		0.005	0.046	0.087	0.136	0.232	0.273	0.232	0.136	0.087	0.046	0.005	
	5			0.009	0.023	0.047	0.124	0.219	0.279	0.254	0.208	0.147	0.033	0.005
	6			0.001	0.004	0.010	0.041	0.109	0.209	0.296	0.312	0.294	0.149	0.051
	7					0.001	0.008	0.031	0.090	0.198	0.267	0.336	0.383	0.279
	8						0.001	0.004	0.017	0.058	0.100	0.168	0.430	0.663

TABLE III BINOMIAL PROBABILITIES *(continued)*

n	x	0.05	0.1	0.2	0.25	0.3	0.4	0.5	0.6	0.7	0.75	0.8	0.9	0.95
9	0	0.630	0.387	0.134	0.075	0.040	0.010	0.002						
	1	0.299	0.387	0.302	0.225	0.156	0.060	0.018	0.004					
	2	0.063	0.172	0.302	0.300	0.267	0.161	0.070	0.021	0.004	0.001			
	3	0.008	0.045	0.176	0.234	0.267	0.251	0.164	0.074	0.021	0.009	0.003		
	4	0.001	0.007	0.066	0.117	0.172	0.251	0.246	0.167	0.074	0.039	0.017	0.001	
	5		0.001	0.017	0.039	0.074	0.167	0.246	0.251	0.172	0.117	0.066	0.007	0.001
	6			0.003	0.009	0.021	0.074	0.164	0.251	0.267	0.234	0.176	0.045	0.008
	7				0.001	0.004	0.021	0.070	0.161	0.267	0.300	0.302	0.172	0.063
	8						0.004	0.018	0.060	0.156	0.225	0.302	0.387	0.299
	9							0.002	0.010	0.040	0.075	0.134	0.387	0.630
10	0	0.599	0.349	0.107	0.056	0.028	0.006	0.001						
	1	0.315	0.387	0.268	0.188	0.121	0.040	0.010	0.002					
	2	0.075	0.194	0.302	0.282	0.233	0.121	0.044	0.011	0.001				
	3	0.010	0.057	0.201	0.250	0.267	0.215	0.117	0.042	0.009	0.003	0.001		
	4	0.001	0.011	0.088	0.146	0.200	0.251	0.205	0.111	0.037	0.016	0.006		
	5		0.001	0.026	0.058	0.103	0.201	0.246	0.201	0.103	0.058	0.026	0.001	
	6			0.006	0.016	0.037	0.111	0.205	0.251	0.200	0.146	0.088	0.011	0.001
	7			0.001	0.003	0.009	0.042	0.117	0.215	0.267	0.250	0.201	0.057	0.010
	8					0.001	0.011	0.044	0.121	0.233	0.282	0.302	0.194	0.075
	9						0.002	0.010	0.040	0.121	0.188	0.268	0.387	0.315
	10							0.001	0.006	0.028	0.056	0.107	0.349	0.599
11	0	0.569	0.314	0.086	0.042	0.020	0.004							
	1	0.329	0.384	0.236	0.155	0.093	0.027	0.005	0.001					
	2	0.087	0.213	0.295	0.258	0.200	0.089	0.027	0.005	0.001				
	3	0.014	0.071	0.221	0.258	0.257	0.177	0.081	0.023	0.004	0.001			
	4	0.001	0.016	0.111	0.172	0.220	0.236	0.161	0.070	0.017	0.006	0.002		
	5		0.002	0.039	0.080	0.132	0.221	0.226	0.147	0.057	0.027	0.010		
	6			0.010	0.027	0.057	0.147	0.226	0.221	0.132	0.080	0.039	0.002	
	7			0.002	0.006	0.017	0.070	0.161	0.236	0.220	0.172	0.111	0.016	0.001
	8				0.001	0.004	0.023	0.081	0.177	0.257	0.258	0.221	0.071	0.014
	9					0.001	0.005	0.027	0.089	0.200	0.258	0.295	0.213	0.087
	10						0.001	0.005	0.027	0.093	0.155	0.236	0.384	0.329
	11								0.004	0.020	0.042	0.086	0.314	0.569
12	0	0.540	0.282	0.069	0.032	0.014	0.002							
	1	0.341	0.377	0.206	0.127	0.071	0.017	0.003						
	2	0.099	0.230	0.283	0.232	0.168	0.064	0.016	0.002					
	3	0.017	0.085	0.236	0.258	0.240	0.142	0.054	0.012	0.001				
	4	0.002	0.021	0.133	0.194	0.231	0.213	0.121	0.042	0.008	0.002	0.001		
	5		0.004	0.053	0.103	0.158	0.227	0.193	0.101	0.029	0.012	0.003		
	6			0.016	0.040	0.079	0.177	0.226	0.177	0.079	0.040	0.016		
	7			0.003	0.012	0.029	0.101	0.193	0.227	0.158	0.103	0.053	0.004	
	8			0.001	0.002	0.008	0.042	0.121	0.213	0.231	0.194	0.133	0.021	0.002
	9					0.001	0.012	0.054	0.142	0.240	0.258	0.236	0.085	0.017
	10						0.002	0.016	0.064	0.168	0.232	0.283	0.230	0.099
	11							0.003	0.017	0.071	0.127	0.206	0.377	0.341
	12								0.002	0.014	0.032	0.069	0.282	0.540

TABLE III   BINOMIAL PROBABILITIES *(continued)*

| | | | | | | | | P | | | | | | |
n	x	0.05	0.1	0.2	0.25	0.3	0.4	0.5	0.6	0.7	0.75	0.8	0.9	0.95
13	0	0.513	0.254	0.055	0.024	0.010	0.001							
	1	0.351	0.367	0.179	0.103	0.054	0.011	0.002						
	2	0.111	0.245	0.268	0.206	0.139	0.045	0.010	0.001					
	3	0.021	0.100	0.246	0.252	0.218	0.111	0.035	0.006	0.001				
	4	0.003	0.028	0.154	0.210	0.234	0.184	0.087	0.024	0.003				
	5		0.006	0.069	0.126	0.180	0.221	0.157	0.066	0.014	0.005	0.001		
	6		0.001	0.023	0.056	0.103	0.197	0.209	0.131	0.044	0.019	0.006		
	7			0.006	0.019	0.044	0.131	0.209	0.197	0.103	0.056	0.023	0.001	
	8			0.001	0.005	0.014	0.066	0.157	0.221	0.180	0.126	0.069	0.006	
	9				0.001	0.003	0.024	0.087	0.184	0.234	0.210	0.154	0.028	0.003
	10					0.001	0.006	0.035	0.111	0.218	0.252	0.246	0.100	0.021
	11						0.001	0.010	0.045	0.139	0.206	0.268	0.245	0.111
	12							0.002	0.011	0.054	0.103	0.179	0.367	0.351
	13								0.001	0.010	0.024	0.055	0.254	0.513
14	0	0.488	0.229	0.044	0.018	0.007	0.001							
	1	0.359	0.356	0.154	0.083	0.041	0.007	0.001						
	2	0.123	0.257	0.250	0.180	0.113	0.032	0.006	0.001					
	3	0.026	0.114	0.250	0.240	0.194	0.085	0.022	0.003					
	4	0.004	0.035	0.172	0.220	0.229	0.155	0.061	0.014	0.001				
	5		0.008	0.086	0.147	0.196	0.207	0.122	0.041	0.007	0.002			
	6		0.001	0.032	0.073	0.126	0.207	0.183	0.092	0.023	0.008	0.002		
	7			0.009	0.028	0.062	0.157	0.209	0.157	0.062	0.028	0.009		
	8			0.002	0.008	0.023	0.092	0.183	0.207	0.126	0.073	0.032	0.001	
	9				0.002	0.007	0.041	0.122	0.207	0.196	0.147	0.086	0.008	
	10					0.001	0.014	0.061	0.155	0.229	0.220	0.172	0.035	0.004
	11						0.003	0.022	0.085	0.194	0.240	0.250	0.114	0.026
	12						0.001	0.006	0.032	0.113	0.180	0.250	0.257	0.123
	13							0.001	0.007	0.041	0.083	0.154	0.356	0.359
	14								0.001	0.007	0.018	0.044	0.229	0.488
15	0	0.463	0.206	0.035	0.013	0.005								
	1	0.366	0.343	0.132	0.067	0.031	0.005							
	2	0.135	0.267	0.231	0.156	0.092	0.022	0.003						
	3	0.031	0.129	0.250	0.225	0.170	0.063	0.014	0.002					
	4	0.005	0.043	0.188	0.225	0.219	0.127	0.042	0.007	0.001				
	5	0.001	0.010	0.103	0.165	0.206	0.186	0.092	0.024	0.003	0.001			
	6		0.002	0.043	0.092	0.147	0.207	0.153	0.061	0.012	0.003	0.001		
	7			0.014	0.039	0.081	0.177	0.196	0.118	0.035	0.013	0.003		
	8			0.003	0.013	0.035	0.118	0.196	0.177	0.081	0.039	0.014		
	9			0.001	0.003	0.012	0.061	0.153	0.207	0.147	0.092	0.043	0.002	
	10				0.001	0.003	0.024	0.092	0.186	0.206	0.165	0.103	0.010	0.001
	11					0.001	0.007	0.042	0.127	0.219	0.225	0.188	0.043	0.005
	12						0.002	0.014	0.063	0.170	0.225	0.250	0.129	0.031
	13							0.003	0.022	0.092	0.156	0.231	0.267	0.135
	14								0.005	0.031	0.067	0.132	0.343	0.366
	15									0.005	0.013	0.035	0.206	0.463

TABLE IV   AREAS UNDER THE NORMAL CURVE

An entry in the table is the proportion under the entire curve which is between $z = 0$ and a positive value of z. Areas for negative values of z are obtained by symmetry.

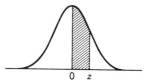

Second decimal place of z

z	.00	.01	.02	.03	.04	.05	.06	.07	.08	.09
0.0	.0000	.0040	.0080	.0120	.0160	.0199	.0239	.0279	.0319	.0359
0.1	.0398	.0438	.0478	.0517	.0557	.0596	.0636	.0675	.0714	.0753
0.2	.0793	.0832	.0871	.0910	.0948	.0987	.1026	.1064	.1103	.1141
0.3	.1179	.1217	.1255	.1293	.1331	.1368	.1406	.1443	.1480	.1517
0.4	.1554	.1591	.1628	.1664	.1700	.1736	.1772	.1808	.1844	.1879
0.5	.1915	.1950	.1985	.2019	.2054	.2088	.2123	.2157	.2190	.2224
0.6	.2257	.2291	.2324	.2357	.2389	.2422	.2454	.2486	.2517	.2549
0.7	.2580	.2611	.2642	.2673	.2703	.2734	.2764	.2794	.2823	.2852
0.8	.2881	.2910	.2939	.2967	.2995	.3023	.3051	.3078	.3106	.3133
0.9	.3159	.3186	.3212	.3238	.3264	.3289	.3315	.3340	.3365	.3389
1.0	.3413	.3438	.3461	.3485	.3508	.3531	.3554	.3577	.3599	.3621
1.1	.3643	.3665	.3686	.3708	.3729	.3749	.3770	.3790	.3810	.3830
1.2	.3849	.3869	.3888	.3907	.3925	.3944	.3962	.3980	.3997	.4015
1.3	.4032	.4049	.4066	.4082	.4099	.4115	.4131	.4147	.4162	.4177
1.4	.4192	.4207	.4222	.4236	.4251	.4265	.4279	.4292	.4306	.4319
1.5	.4332	.4345	.4357	.4370	.4382	.4394	.4406	.4418	.4429	.4441
1.6	.4452	.4463	.4474	.4484	.4495	.4505	.4515	.4525	.4535	.4545
1.7	.4554	.4564	.4573	.4582	.4591	.4599	.4608	.4616	.4625	.4633
1.8	.4641	.4649	.4656	.4664	.4671	.4678	.4686	.4693	.4699	.4706
1.9	.4713	.4719	.4726	.4732	.4738	.4744	.4750	.4756	.4761	.4767
2.0	.4772	.4778	.4783	.4788	.4793	.4798	.4803	.4808	.4812	.4817
2.1	.4821	.4826	.4830	.4834	.4838	.4842	.4846	.4850	.4854	.4857
2.2	.4861	.4864	.4868	.4871	.4875	.4878	.4881	.4884	.4887	.4890
2.3	.4893	.4896	.4898	.4901	.4904	.4906	.4909	.4911	.4913	.4916
2.4	.4918	.4920	.4922	.4925	.4927	.4929	.4931	.4932	.4934	.4936
2.5	.4938	.4940	.4941	.4943	.4945	.4946	.4948	.4949	.4951	.4952
2.6	.4953	.4955	.4956	.4957	.4959	.4960	.4961	.4962	.4963	.4964
2.7	.4965	.4966	.4967	.4968	.4969	.4970	.4971	.4972	.4973	.4974
2.8	.4974	.4975	.4976	.4977	.4977	.4978	.4979	.4979	.4980	.4981
2.9	.4981	.4982	.4982	.4983	.4984	.4984	.4985	.4985	.4986	.4986
3.0	.4987	.4987	.4987	.4988	.4988	.4989	.4989	.4989	.4990	.4990

From Paul G. Hoel, *Elementary Statistics*, 3rd ed., © 1971, John Wiley and Sons, Inc., New York, p. 287.

TABLE V   CRITICAL POINTS OF THE t DISTRIBUTION

The first column lists the number of degrees of freedom (ν). The headings of the other columns give probabilities (P) for t to exceed the entry value. Use symmetry for negative t values.

P df	.10	.05	.025	.01	.005
1	3.078	6.314	12.706	31.821	63.657
2	1.886	2.920	4.303	6.965	9.925
3	1.638	2.353	3.182	4.541	5.841
4	1.533	2.132	2.776	3.747	4.604
5	1.476	2.015	2.571	3.365	4.032
6	1.440	1.943	2.447	3.143	3.707
7	1.415	1.895	2.365	2.998	3.499
8	1.397	1.860	2.306	2.896	3.355
9	1.383	1.833	2.262	2.821	3.250
10	1.372	1.812	2.228	2.764	3.169
11	1.363	1.796	2.201	2.718	3.106
12	1.356	1.782	2.179	2.681	3.055
13	1.350	1.771	2.160	2.650	3.012
14	1.345	1.761	2.145	2.624	2.977
15	1.341	1.753	2.131	2.602	2.947
16	1.337	1.746	2.120	2.583	2.921
17	1.333	1.740	2.110	2.567	2.898
18	1.330	1.734	2.101	2.552	2.878
19	1.328	1.729	2.093	2.539	2.861
20	1.325	1.725	2.086	2.528	2.845
21	1.323	1.721	2.080	2.518	2.831
22	1.321	1.717	2.074	2.508	2.819
23	1.319	1.714	2.069	2.500	2.807
24	1.318	1.711	2.064	2.492	2.797
25	1.316	1.708	2.060	2.485	2.787
26	1.315	1.706	2.056	2.479	2.779
27	1.314	1.703	2.052	2.473	2.771
28	1.313	1.701	2.048	2.467	2.763
29	1.311	1.699	2.045	2.462	2.756
30	1.310	1.697	2.042	2.457	2.750
40	1.303	1.684	2.021	2.423	2.704
60	1.296	1.671	2.000	2.390	2.660
120	1.289	1.658	1.980	2.358	2.617
∞	1.282	1.645	1.960	2.326	2.576

From Paul G. Hoel, *Elementary Statistics,* 3rd ed., © 1971, John Wiley and Sons, Inc., New York, p. 288.

TABLE VI CRITICAL POINTS OF THE *F* DISTRIBUTION

5% (Roman Type) and 1% (Boldface Type) Points for the Distribution of *F*

Degrees of freedom for numerator (df_1)

Degrees of freedom for denominator (df_2)	1	2	3	4	5	6	7	8	9	10	11	12	14	16	20	24	30	40	50	75	100	200	500	∞
1	161 / 4052	200 / 4999	216 / 5403	225 / 5625	230 / 5764	234 / 5859	237 / 5928	239 / 5981	241 / 6022	242 / 6056	243 / 6082	244 / 6106	245 / 6142	246 / 6169	248 / 6208	249 / 6234	250 / 6258	251 / 6286	252 / 6302	253 / 6323	253 / 6334	254 / 6352	254 / 6361	254 / 6366
2	18.51 / 98.49	19.00 / 99.01	19.16 / 99.17	19.25 / 99.25	19.30 / 99.30	19.33 / 99.33	19.36 / 99.34	19.37 / 99.36	19.38 / 99.38	19.39 / 99.40	19.40 / 99.41	19.41 / 99.42	19.42 / 99.43	19.43 / 99.44	19.44 / 99.45	19.45 / 99.46	19.46 / 99.47	19.47 / 99.48	19.47 / 99.48	19.48 / 99.49	19.49 / 99.49	19.49 / 99.49	19.50 / 99.50	19.50 / 99.50
3	10.13 / 34.12	9.55 / 30.81	9.28 / 29.46	9.12 / 28.71	9.01 / 28.24	8.94 / 27.91	8.88 / 27.67	8.84 / 27.49	8.81 / 27.34	8.78 / 27.23	8.76 / 27.13	8.74 / 27.05	8.71 / 26.92	8.69 / 26.83	8.66 / 26.69	8.64 / 26.60	8.62 / 26.50	8.60 / 26.41	8.58 / 26.30	8.57 / 26.27	8.56 / 26.23	8.54 / 26.18	8.54 / 26.14	8.53 / 26.12
4	7.71 / 21.20	6.94 / 18.00	6.59 / 16.69	6.39 / 15.98	6.26 / 15.52	6.16 / 15.21	6.09 / 14.98	6.04 / 14.80	6.00 / 14.66	5.96 / 14.54	5.93 / 14.45	5.91 / 14.37	5.87 / 14.24	5.84 / 14.15	5.80 / 14.02	5.77 / 13.93	5.74 / 13.83	5.71 / 13.74	5.70 / 13.69	5.68 / 13.61	5.66 / 13.57	5.65 / 13.52	5.64 / 13.48	5.63 / 13.46
5	6.61 / 16.26	5.79 / 13.27	5.41 / 12.06	5.19 / 11.39	5.05 / 10.97	4.95 / 10.67	4.88 / 10.45	4.82 / 10.27	4.78 / 10.15	4.74 / 10.05	4.70 / 9.96	4.68 / 9.89	4.64 / 9.77	4.60 / 9.68	4.56 / 9.55	4.53 / 9.47	4.50 / 9.38	4.46 / 9.29	4.44 / 9.24	4.42 / 9.17	4.40 / 9.13	4.38 / 9.07	4.37 / 9.04	4.36 / 9.02
6	5.99 / 13.74	5.14 / 10.92	4.76 / 9.78	4.53 / 9.15	4.39 / 8.75	4.28 / 8.47	4.21 / 8.26	4.15 / 8.10	4.10 / 7.98	4.06 / 7.87	4.03 / 7.79	4.00 / 7.72	3.96 / 7.60	3.92 / 7.52	3.87 / 7.39	3.84 / 7.31	3.81 / 7.23	3.77 / 7.14	3.75 / 7.09	3.72 / 7.02	3.71 / 6.99	3.69 / 6.94	3.68 / 6.90	3.67 / 6.88
7	5.59 / 12.25	4.74 / 9.55	4.35 / 8.45	4.12 / 7.85	3.97 / 7.46	3.87 / 7.19	3.79 / 7.00	3.73 / 6.84	3.68 / 6.71	3.63 / 6.62	3.60 / 6.54	3.57 / 6.47	3.52 / 6.35	3.49 / 6.27	3.44 / 6.15	3.41 / 6.07	3.38 / 5.98	3.34 / 5.90	3.32 / 5.85	3.29 / 5.78	3.28 / 5.75	3.25 / 5.70	3.24 / 5.67	3.23 / 5.65
8	5.32 / 11.26	4.46 / 8.65	4.07 / 7.59	3.84 / 7.01	3.69 / 6.63	3.58 / 6.37	3.50 / 6.19	3.44 / 6.03	3.39 / 5.91	3.34 / 5.82	3.31 / 5.74	3.28 / 5.67	3.23 / 5.56	3.20 / 5.48	3.15 / 5.36	3.12 / 5.28	3.08 / 5.20	3.05 / 5.11	3.03 / 5.06	3.00 / 5.00	2.98 / 4.96	2.96 / 4.91	2.94 / 4.88	2.93 / 4.86
9	5.12 / 10.56	4.26 / 8.02	3.86 / 6.99	3.63 / 6.42	3.48 / 6.06	3.37 / 5.80	3.29 / 5.62	3.23 / 5.47	3.18 / 5.35	3.13 / 5.26	3.10 / 5.18	3.07 / 5.11	3.02 / 5.00	2.98 / 4.92	2.93 / 4.80	2.90 / 4.73	2.86 / 4.64	2.82 / 4.56	2.80 / 4.51	2.77 / 4.45	2.76 / 4.41	2.73 / 4.36	2.72 / 4.33	2.71 / 4.31

TABLE VI (continued)

Degrees of freedom for numerator (df_1)

Degrees of freedom for denominator (df_2)	1	2	3	4	5	6	7	8	9	10	11	12	14	16	20	24	30	40	50	75	100	200	500	∞
10	4.96 10.04	4.10 7.56	3.71 6.55	3.48 5.99	3.33 5.64	3.22 5.39	3.14 5.21	3.07 5.06	3.02 4.95	2.97 4.85	2.94 4.78	2.91 4.71	2.86 4.60	2.82 4.52	2.77 4.41	2.74 4.33	2.70 4.25	2.67 4.17	2.64 4.12	2.61 4.05	2.59 4.01	2.56 3.96	2.55 3.93	2.54 3.91
11	4.84 9.65	3.98 7.20	3.59 6.22	3.36 5.67	3.20 5.32	3.09 5.07	3.01 4.88	2.95 4.74	2.90 4.63	2.86 4.54	2.82 4.46	2.79 4.40	2.74 4.29	2.70 4.21	2.65 4.10	2.61 4.02	2.57 3.94	2.53 3.86	2.50 3.80	2.47 3.74	2.45 3.70	2.42 3.66	2.41 3.62	2.40 3.60
12	4.75 9.33	3.88 6.93	3.49 5.95	3.26 5.41	3.11 5.06	3.00 4.82	2.92 4.65	2.85 4.50	2.80 4.39	2.76 4.30	2.72 4.22	2.69 4.16	2.64 4.05	2.60 3.98	2.54 3.86	2.50 3.78	2.46 3.70	2.42 3.61	2.40 3.56	2.36 3.49	2.35 3.46	2.32 3.41	2.31 3.38	2.30 3.36
13	4.67 9.07	3.80 6.70	3.41 5.74	3.18 5.20	3.02 4.86	2.92 4.62	2.84 4.44	2.77 4.30	2.72 4.19	2.67 4.10	2.63 4.02	2.60 3.96	2.55 3.85	2.51 3.78	2.46 3.67	2.42 3.59	2.38 3.51	2.34 3.42	2.32 3.37	2.28 3.30	2.26 3.27	2.24 3.21	2.22 3.18	2.21 3.16
14	4.60 8.86	3.74 6.51	3.34 5.56	3.11 5.03	2.96 4.69	2.85 4.46	2.77 4.28	2.70 4.14	2.65 4.03	2.60 3.94	2.56 3.86	2.53 3.80	2.48 3.70	2.44 3.62	2.39 3.51	2.35 3.43	2.31 3.34	2.27 3.26	2.24 3.21	2.21 3.14	2.19 3.11	2.16 3.06	2.14 3.02	2.13 3.00
15	4.54 8.68	3.68 6.36	3.29 5.42	3.06 4.89	2.90 4.56	2.79 4.32	2.70 4.14	2.64 4.00	2.59 3.89	2.55 3.80	2.51 3.73	2.48 3.67	2.43 3.56	2.39 3.48	2.33 3.36	2.29 3.29	2.25 3.20	2.21 3.12	2.18 3.07	2.15 3.00	2.12 2.97	2.10 2.92	2.08 2.89	2.07 2.87
16	4.49 8.53	3.63 6.23	3.24 5.29	3.01 4.77	2.85 4.44	2.74 4.20	2.66 4.03	2.59 3.89	2.54 3.78	2.49 3.69	2.45 3.61	2.42 3.55	2.37 3.45	2.33 3.37	2.28 3.25	2.24 3.18	2.20 3.10	2.16 3.01	2.13 2.96	2.09 2.89	2.07 2.86	2.04 2.80	2.02 2.77	2.01 2.75
17	4.45 8.40	3.59 6.11	3.20 5.18	2.96 4.67	2.81 4.34	2.70 4.10	2.62 3.93	2.55 3.79	2.50 3.68	2.45 3.59	2.41 3.52	2.38 3.45	2.33 3.35	2.29 3.27	2.23 3.16	2.19 3.08	2.15 3.00	2.11 2.92	2.08 2.86	2.04 2.79	2.02 2.76	1.99 2.70	1.97 2.67	1.96 2.65
18	4.41 8.28	3.55 6.01	3.16 5.09	2.93 4.58	2.77 4.25	2.66 4.01	2.58 3.85	2.51 3.71	2.46 3.60	2.41 3.51	2.37 3.44	2.34 3.37	2.29 3.27	2.25 3.19	2.19 3.07	2.15 3.00	2.11 2.91	2.07 2.83	2.04 2.78	2.00 2.71	1.98 2.68	1.95 2.62	1.93 2.59	1.92 2.57
19	4.38 8.18	3.52 5.93	3.13 5.01	2.90 4.50	2.74 4.17	2.63 3.94	2.55 3.77	2.48 3.63	2.43 3.52	2.38 3.43	2.34 3.36	2.31 3.30	2.26 3.19	2.21 3.12	2.15 3.00	2.11 2.92	2.07 2.84	2.02 2.76	2.00 2.70	1.96 2.63	1.94 2.60	1.91 2.54	1.90 2.51	1.88 2.49
20	4.35 8.10	3.49 5.85	3.10 4.94	2.87 4.43	2.71 4.10	2.60 3.87	2.52 3.71	2.45 3.56	2.40 3.45	2.35 3.37	2.31 3.30	2.28 3.23	2.23 3.13	2.18 3.05	2.12 2.94	2.08 2.86	2.04 2.77	1.99 2.69	1.96 2.63	1.92 2.56	1.90 2.53	1.87 2.47	1.85 2.44	1.84 2.42
21	4.32 8.02	3.47 5.78	3.07 4.87	2.84 4.37	2.68 4.04	2.57 3.81	2.49 3.65	2.42 3.51	2.37 3.40	2.32 3.31	2.28 3.24	2.25 3.17	2.20 3.07	2.15 2.99	2.09 2.88	2.05 2.80	2.00 2.72	1.96 2.63	1.93 2.58	1.89 2.51	1.87 2.47	1.84 2.42	1.82 2.38	1.81 2.36
22	4.30 7.94	3.44 5.72	3.05 4.82	2.82 4.31	2.66 3.99	2.55 3.76	2.47 3.59	2.40 3.45	2.35 3.35	2.30 3.26	2.26 3.18	2.23 3.12	2.18 3.02	2.13 2.94	2.07 2.83	2.03 2.75	1.98 2.67	1.93 2.58	1.91 2.53	1.87 2.46	1.84 2.42	1.81 2.37	1.80 2.33	1.78 2.31
23	4.28 7.88	3.42 5.66	3.03 4.76	2.80 4.26	2.64 3.94	2.53 3.71	2.45 3.54	2.38 3.41	2.32 3.30	2.28 3.21	2.24 3.14	2.20 3.07	2.14 2.97	2.10 2.89	2.04 2.78	2.00 2.70	1.96 2.62	1.91 2.53	1.88 2.48	1.84 2.41	1.82 2.37	1.79 2.32	1.77 2.28	1.76 2.26

24	1.73 / 2.21	1.74 / 2.23	1.76 / 2.27	1.80 / 2.33	1.82 / 2.36	1.86 / 2.44	1.89 / 2.49	1.94 / 2.58	1.98 / 2.66	2.02 / 2.74	2.09 / 2.85	2.13 / 2.93	2.18 / 3.03	2.22 / 3.09	2.26 / 3.17	2.30 / 3.25	2.36 / 3.36	2.43 / 3.50	2.51 / 3.67	2.62 / 3.90	2.78 / 4.22	3.01 / 4.72	3.40 / 5.61	4.26 / 7.82
25	1.71 / 2.17	1.72 / 2.19	1.74 / 2.23	1.77 / 2.29	1.80 / 2.32	1.84 / 2.40	1.87 / 2.45	1.92 / 2.54	1.96 / 2.62	2.00 / 2.70	2.06 / 2.81	2.11 / 2.89	2.16 / 2.99	2.20 / 3.05	2.24 / 3.13	2.28 / 3.21	2.34 / 3.32	2.41 / 3.46	2.49 / 3.63	2.60 / 3.86	2.76 / 4.18	2.99 / 4.68	3.38 / 5.57	4.24 / 7.77
26	1.69 / 2.13	1.70 / 2.15	1.72 / 2.19	1.76 / 2.25	1.78 / 2.28	1.82 / 2.36	1.85 / 2.41	1.90 / 2.50	1.95 / 2.58	1.99 / 2.66	2.05 / 2.77	2.10 / 2.86	2.15 / 2.96	2.18 / 3.02	2.22 / 3.09	2.27 / 3.17	2.32 / 3.29	2.39 / 3.42	2.47 / 3.59	2.59 / 3.82	2.74 / 4.14	2.96 / 4.64	3.37 / 5.53	4.22 / 7.72
27	1.67 / 2.10	1.68 / 2.12	1.71 / 2.16	1.74 / 2.21	1.76 / 2.25	1.80 / 2.33	1.84 / 2.38	1.88 / 2.47	1.93 / 2.55	1.97 / 2.63	2.03 / 2.74	2.08 / 2.83	2.13 / 2.93	2.16 / 2.98	2.20 / 3.06	2.25 / 3.14	2.30 / 3.26	2.37 / 3.39	2.46 / 3.56	2.57 / 3.79	2.73 / 4.11	2.96 / 4.60	3.35 / 5.49	4.21 / 7.68
28	1.65 / 2.06	1.67 / 2.09	1.69 / 2.13	1.72 / 2.18	1.75 / 2.22	1.78 / 2.30	1.81 / 2.35	1.87 / 2.44	1.91 / 2.52	1.96 / 2.60	2.02 / 2.71	2.06 / 2.80	2.12 / 2.90	2.15 / 2.95	2.19 / 3.03	2.24 / 3.11	2.29 / 3.23	2.36 / 3.36	2.44 / 3.53	2.56 / 3.76	2.71 / 4.07	2.95 / 4.57	3.34 / 5.45	4.20 / 7.64
29	1.64 / 2.03	1.65 / 2.06	1.68 / 2.10	1.71 / 2.15	1.73 / 2.19	1.77 / 2.27	1.80 / 2.32	1.85 / 2.41	1.90 / 2.49	1.94 / 2.57	2.00 / 2.68	2.05 / 2.77	2.10 / 2.87	2.14 / 2.92	2.18 / 3.00	2.22 / 3.08	2.28 / 3.20	2.35 / 3.33	2.43 / 3.50	2.54 / 3.73	2.70 / 4.04	2.93 / 4.54	3.33 / 5.42	4.18 / 7.60
30	1.62 / 2.01	1.64 / 2.03	1.66 / 2.07	1.69 / 2.13	1.72 / 2.16	1.76 / 2.24	1.79 / 2.29	1.84 / 2.38	1.89 / 2.47	1.93 / 2.55	1.99 / 2.66	2.04 / 2.74	2.09 / 2.84	2.12 / 2.90	2.16 / 2.98	2.21 / 3.06	2.27 / 3.17	2.34 / 3.30	2.42 / 3.47	2.53 / 3.70	2.69 / 4.02	2.92 / 4.51	3.32 / 5.39	4.17 / 7.56
32	1.59 / 1.96	1.61 / 1.98	1.64 / 2.02	1.67 / 2.08	1.69 / 2.12	1.74 / 2.20	1.76 / 2.25	1.82 / 2.34	1.86 / 2.42	1.91 / 2.51	1.97 / 2.62	2.02 / 2.70	2.07 / 2.80	2.10 / 2.86	2.14 / 2.94	2.19 / 3.01	2.25 / 3.12	2.32 / 3.25	2.40 / 3.42	2.51 / 3.66	2.67 / 3.97	2.90 / 4.46	3.30 / 5.34	4.15 / 7.50
34	1.57 / 1.91	1.59 / 1.94	1.61 / 1.98	1.64 / 2.04	1.67 / 2.08	1.71 / 2.15	1.74 / 2.21	1.80 / 2.30	1.84 / 2.38	1.89 / 2.47	1.95 / 2.58	2.00 / 2.66	2.05 / 2.76	2.08 / 2.82	2.12 / 2.89	2.17 / 2.97	2.23 / 3.08	2.30 / 3.21	2.38 / 3.38	2.49 / 3.61	2.65 / 3.93	2.88 / 4.42	3.28 / 5.29	4.13 / 7.44
36	1.55 / 1.87	1.56 / 1.90	1.59 / 1.94	1.62 / 2.00	1.65 / 2.04	1.69 / 2.12	1.72 / 2.17	1.78 / 2.26	1.82 / 2.35	1.87 / 2.43	1.93 / 2.54	1.98 / 2.62	2.03 / 2.72	2.06 / 2.78	2.10 / 2.86	2.15 / 2.94	2.21 / 3.04	2.28 / 3.18	2.36 / 3.35	2.48 / 3.58	2.63 / 3.89	2.86 / 4.38	3.26 / 5.25	4.11 / 7.39
38	1.53 / 1.84	1.54 / 1.86	1.57 / 1.90	1.60 / 1.97	1.63 / 2.00	1.67 / 2.08	1.71 / 2.14	1.76 / 2.22	1.80 / 2.32	1.85 / 2.40	1.92 / 2.51	1.96 / 2.59	2.02 / 2.69	2.05 / 2.75	2.09 / 2.82	2.14 / 2.91	2.19 / 3.02	2.26 / 3.15	2.35 / 3.32	2.46 / 3.54	2.62 / 3.86	2.85 / 4.34	3.25 / 5.21	4.10 / 7.35
40	1.51 / 1.81	1.53 / 1.84	1.55 / 1.88	1.59 / 1.94	1.61 / 1.97	1.66 / 2.05	1.69 / 2.11	1.74 / 2.20	1.79 / 2.29	1.84 / 2.37	1.90 / 2.49	1.95 / 2.56	2.00 / 2.66	2.04 / 2.73	2.07 / 2.80	2.12 / 2.88	2.18 / 2.99	2.25 / 3.12	2.34 / 3.29	2.45 / 3.51	2.61 / 3.83	2.84 / 4.31	3.23 / 5.18	4.08 / 7.31
42	1.49 / 1.78	1.51 / 1.80	1.54 / 1.85	1.57 / 1.91	1.60 / 1.94	1.64 / 2.02	1.68 / 2.08	1.73 / 2.17	1.78 / 2.26	1.82 / 2.35	1.89 / 2.46	1.94 / 2.54	1.99 / 2.64	2.02 / 2.70	2.06 / 2.77	2.11 / 2.86	2.17 / 2.96	2.24 / 3.10	2.32 / 3.26	2.44 / 3.49	2.59 / 3.80	2.83 / 4.29	3.22 / 5.15	4.07 / 7.27
44	1.48 / 1.75	1.50 / 1.78	1.52 / 1.82	1.56 / 1.88	1.58 / 1.92	1.63 / 2.00	1.66 / 2.06	1.72 / 2.15	1.76 / 2.24	1.81 / 2.32	1.88 / 2.44	1.92 / 2.52	1.98 / 2.62	2.01 / 2.68	2.05 / 2.75	2.10 / 2.84	2.16 / 2.94	2.23 / 3.07	2.31 / 3.24	2.43 / 3.46	2.58 / 3.78	2.82 / 4.26	3.21 / 5.12	4.06 / 7.24
46	1.46 / 1.72	1.48 / 1.76	1.51 / 1.80	1.54 / 1.86	1.57 / 1.90	1.62 / 1.98	1.65 / 2.04	1.71 / 2.13	1.75 / 2.22	1.80 / 2.30	1.87 / 2.42	1.91 / 2.50	1.97 / 2.60	2.00 / 2.66	2.04 / 2.73	2.09 / 2.82	2.14 / 2.92	2.22 / 3.05	2.30 / 3.22	2.42 / 3.44	2.57 / 3.76	2.81 / 4.24	3.20 / 5.10	4.05 / 7.21
48	1.45 / 1.70	1.47 / 1.73	1.50 / 1.78	1.53 / 1.84	1.56 / 1.88	1.61 / 1.96	1.64 / 2.02	1.70 / 2.11	1.74 / 2.20	1.79 / 2.28	1.86 / 2.40	1.90 / 2.48	1.96 / 2.58	1.99 / 2.64	2.03 / 2.71	2.08 / 2.80	2.14 / 2.90	2.21 / 3.04	2.30 / 3.20	2.41 / 3.42	2.56 / 3.74	2.80 / 4.22	3.19 / 5.08	4.04 / 7.19

TABLE VI (continued)

Each cell shows the upper value (top) over the lower value (bottom).

df₂ \ df₁	1	2	3	4	5	6	7	8	9	10	11	12	14	16	20	24	30	40	50	75	100	200	500	∞
50	4.03 / 7.17	3.18 / 5.06	2.79 / 4.20	2.56 / 3.72	2.40 / 3.41	2.29 / 3.18	2.20 / 3.02	2.13 / 2.88	2.07 / 2.78	2.02 / 2.70	1.98 / 2.62	1.95 / 2.56	1.90 / 2.46	1.85 / 2.39	1.78 / 2.26	1.74 / 2.18	1.69 / 2.10	1.63 / 2.00	1.60 / 1.94	1.55 / 1.86	1.52 / 1.82	1.48 / 1.76	1.46 / 1.71	1.44 / 1.68
55	4.02 / 7.12	3.17 / 5.01	2.78 / 4.16	2.54 / 3.68	2.38 / 3.37	2.27 / 3.15	2.18 / 2.98	2.11 / 2.85	2.05 / 2.75	2.00 / 2.66	1.97 / 2.59	1.93 / 2.53	1.88 / 2.43	1.83 / 2.35	1.76 / 2.23	1.72 / 2.15	1.67 / 2.06	1.61 / 1.96	1.58 / 1.90	1.52 / 1.82	1.50 / 1.78	1.46 / 1.71	1.43 / 1.66	1.41 / 1.64
60	4.00 / 7.08	3.15 / 4.98	2.76 / 4.13	2.52 / 3.65	2.37 / 3.34	2.25 / 3.12	2.17 / 2.95	2.10 / 2.82	2.04 / 2.72	1.99 / 2.63	1.95 / 2.56	1.92 / 2.50	1.86 / 2.40	1.81 / 2.32	1.75 / 2.20	1.70 / 2.12	1.65 / 2.03	1.59 / 1.93	1.56 / 1.87	1.50 / 1.79	1.48 / 1.74	1.44 / 1.68	1.41 / 1.63	1.39 / 1.60
65	3.99 / 7.04	3.14 / 4.95	2.75 / 4.10	2.51 / 3.62	2.36 / 3.31	2.24 / 3.09	2.15 / 2.93	2.08 / 2.79	2.02 / 2.70	1.98 / 2.61	1.94 / 2.54	1.90 / 2.47	1.85 / 2.37	1.80 / 2.30	1.73 / 2.18	1.68 / 2.09	1.63 / 2.00	1.57 / 1.90	1.54 / 1.84	1.49 / 1.76	1.46 / 1.71	1.42 / 1.64	1.39 / 1.60	1.37 / 1.56
70	3.98 / 7.01	3.13 / 4.92	2.74 / 4.08	2.50 / 3.60	2.35 / 3.29	2.23 / 3.07	2.14 / 2.91	2.07 / 2.77	2.01 / 2.67	1.97 / 2.59	1.93 / 2.51	1.89 / 2.45	1.84 / 2.35	1.79 / 2.28	1.72 / 2.15	1.67 / 2.07	1.62 / 1.98	1.56 / 1.88	1.53 / 1.82	1.47 / 1.74	1.45 / 1.69	1.40 / 1.63	1.37 / 1.56	1.35 / 1.53
80	3.96 / 6.96	3.11 / 4.88	2.72 / 4.04	2.48 / 3.56	2.33 / 3.25	2.21 / 3.04	2.12 / 2.87	2.05 / 2.74	1.99 / 2.64	1.95 / 2.55	1.91 / 2.48	1.88 / 2.41	1.82 / 2.32	1.77 / 2.24	1.70 / 2.11	1.65 / 2.03	1.60 / 1.94	1.54 / 1.84	1.51 / 1.78	1.45 / 1.70	1.42 / 1.65	1.38 / 1.57	1.35 / 1.52	1.32 / 1.49
100	3.94 / 6.90	3.09 / 4.82	2.70 / 3.98	2.46 / 3.51	2.30 / 3.20	2.19 / 2.99	2.10 / 2.82	2.03 / 2.69	1.97 / 2.59	1.92 / 2.51	1.88 / 2.43	1.85 / 2.36	1.79 / 2.26	1.75 / 2.19	1.68 / 2.06	1.63 / 1.98	1.57 / 1.89	1.51 / 1.79	1.48 / 1.73	1.42 / 1.64	1.39 / 1.59	1.34 / 1.51	1.30 / 1.46	1.28 / 1.43
125	3.92 / 6.84	3.07 / 4.78	2.68 / 3.94	2.44 / 3.47	2.29 / 3.17	2.17 / 2.95	2.08 / 2.79	2.01 / 2.65	1.95 / 2.56	1.90 / 2.47	1.86 / 2.40	1.83 / 2.33	1.77 / 2.23	1.72 / 2.15	1.65 / 2.03	1.60 / 1.94	1.55 / 1.85	1.49 / 1.75	1.45 / 1.68	1.39 / 1.59	1.36 / 1.54	1.31 / 1.46	1.27 / 1.40	1.25 / 1.37
150	3.91 / 6.81	3.06 / 4.75	2.67 / 3.91	2.43 / 3.44	2.27 / 3.13	2.16 / 2.92	2.07 / 2.76	2.00 / 2.62	1.94 / 2.53	1.89 / 2.44	1.85 / 2.37	1.82 / 2.30	1.76 / 2.20	1.71 / 2.12	1.64 / 2.00	1.59 / 1.91	1.54 / 1.83	1.47 / 1.72	1.44 / 1.66	1.37 / 1.56	1.34 / 1.51	1.29 / 1.43	1.25 / 1.37	1.22 / 1.33
200	3.89 / 6.76	3.04 / 4.71	2.65 / 3.88	2.41 / 3.41	2.26 / 3.11	2.14 / 2.90	2.05 / 2.73	1.98 / 2.60	1.92 / 2.50	1.87 / 2.41	1.83 / 2.34	1.80 / 2.28	1.74 / 2.17	1.69 / 2.09	1.62 / 1.97	1.57 / 1.88	1.52 / 1.79	1.45 / 1.69	1.42 / 1.62	1.35 / 1.53	1.32 / 1.48	1.26 / 1.39	1.22 / 1.33	1.19 / 1.28
400	3.86 / 6.70	3.02 / 4.66	2.62 / 3.83	2.39 / 3.36	2.23 / 3.06	2.12 / 2.85	2.03 / 2.69	1.96 / 2.55	1.90 / 2.46	1.85 / 2.37	1.81 / 2.29	1.78 / 2.23	1.72 / 2.12	1.67 / 2.04	1.60 / 1.92	1.54 / 1.84	1.49 / 1.74	1.42 / 1.64	1.38 / 1.57	1.32 / 1.47	1.28 / 1.42	1.22 / 1.32	1.16 / 1.24	1.13 / 1.19
1000	3.85 / 6.66	3.00 / 4.62	2.61 / 3.80	2.38 / 3.34	2.22 / 3.04	2.10 / 2.82	2.02 / 2.66	1.95 / 2.53	1.89 / 2.43	1.84 / 2.34	1.80 / 2.26	1.76 / 2.20	1.70 / 2.09	1.65 / 2.01	1.58 / 1.89	1.53 / 1.81	1.47 / 1.71	1.41 / 1.61	1.36 / 1.54	1.30 / 1.44	1.26 / 1.38	1.19 / 1.28	1.13 / 1.19	1.08 / 1.11
∞	3.84 / 6.64	2.99 / 4.60	2.60 / 3.78	2.37 / 3.32	2.21 / 3.02	2.09 / 2.80	2.01 / 2.64	1.94 / 2.51	1.88 / 2.41	1.83 / 2.32	1.79 / 2.24	1.75 / 2.18	1.69 / 2.07	1.64 / 1.99	1.57 / 1.87	1.52 / 1.79	1.46 / 1.69	1.40 / 1.59	1.35 / 1.52	1.28 / 1.41	1.24 / 1.36	1.17 / 1.25	1.11 / 1.15	1.00 / 1.00

Degrees of freedom for numerator (df₁)

Degrees of freedom for denominator (df₂)

TABLE VII   CRITICAL VALUES OF r FOR TESTING $\rho = 0$

For a two-sided test α is twice the value listed at the heading of a column of critical r values; hence for $\alpha = 0.05$ choose the 0.025 column.

n \ α	0.05	0.025	0.010	0.005	n \ α	0.05	0.025	0.010	0.005
5	0.805	0.878	0.934	0.959	17	0.412	0.482	0.558	0.606
6	0.729	0.811	0.882	0.917	18	0.400	0.468	0.542	0.590
7	0.669	0.754	0.833	0.875	19	0.389	0.456	0.528	0.575
8	0.621	0.707	0.789	0.834	20	0.378	0.444	0.516	0.561
9	0.582	0.666	0.750	0.798	25	0.337	0.396	0.462	0.505
10	0.549	0.632	0.716	0.765	30	0.306	0.361	0.423	0.463
11	0.521	0.602	0.685	0.735	40	0.264	0.312	0.366	0.402
12	0.497	0.576	0.658	0.708	50	0.235	0.279	0.328	0.361
13	0.476	0.553	0.634	0.684	60	0.214	0.254	0.300	0.330
14	0.457	0.532	0.612	0.661	80	0.185	0.220	0.260	0.286
15	0.441	0.514	0.592	0.641	100	0.165	0.196	0.232	0.256
16	0.426	0.497	0.574	0.623					

Tables VI and VII are from Paul G. Hoel, *Elementary Statistics,* 3rd ed., © 1971, John Wiley and Sons, Inc., New York, pp. 289, 292-294.

TABLE VIII   CRITICAL POINTS OF THE χ^2 DISTRIBUTION

The first column lists the number of degrees of freedom. The headings of the other columns give probabilities (P) for χ^2 to exceed the entry value.

P	0.050	0.025	0.010	0.005
df				
1	3.84146	5.02389	6.63490	7.87944
2	5.99147	7.37776	9.21034	10.5966
3	7.81473	9.34840	11.3449	12.8381
4	9.48773	11.1433	13.2767	14.8602
5	11.0705	12.8325	15.0863	16.7496
6	12.5916	14.4494	16.8119	18.5476
7	14.0671	16.0128	18.4753	20.2777
8	15.5073	17.5346	20.0902	21.9550
9	16.9190	19.0228	21.6660	23.5893
10	18.3070	20.4831	23.2093	25.1882
11	19.6751	21.9200	24.7250	26.7569
12	21.0261	23.3367	26.2170	28.2995
13	22.3621	24.7356	27.6883	29.8194
14	23.6848	26.1190	29.1413	31.3193
15	24.9958	27.4884	30.5779	32.8013
16	26.2962	28.8454	31.9999	34.2672
17	27.5871	30.1910	33.4087	35.7185
18	28.8693	31.5264	34.8053	37.1564
19	30.1435	32.8523	36.1908	38.5822
20	31.4104	34.1696	37.5662	39.9968
21	32.6705	35.4789	38.9321	41.4010
22	33.9244	36.7807	40.2894	42.7956
23	35.1725	38.0757	41.6384	44.1813
24	36.4151	39.3641	42.9798	45.5585
25	37.6525	40.6465	44.3141	46.9278
26	38.8852	41.9232	45.6417	48.2899
27	40.1133	43.1944	46.9630	49.6449
28	41.3372	44.4607	48.2782	50.9933
29	42.5569	45.7222	49.5879	52.3356
30	43.7729	46.9792	50.8922	53.6720
40	55.7585	59.3417	63.6907	66.7659
50	67.5048	71.4202	76.1539	79.4900
60	79.0819	83.2976	88.3794	91.9517
70	90.5312	95.0231	100.425	104.215
80	101.879	106.629	112.329	116.321
90	113.145	118.136	124.116	128.299
100	124.342	129.561	135.807	140.169

Test

This test will take about an hour to complete. It includes material from all chapters. Make sure you have done all the review problems for each chapter before you take it.

1. Draw a frequency distribution for the following data:

   17, 1, 3, 4, 4, 16, 2, 2, 2, 7, 8, 7, 1, 2, 2, 5, 4, 3, 12, 9, 6, 3, 6, 10, 5, 2, 3, 11, 14, 2, 2,

2. Describe briefly the shape of the frequency distribution in problem 1.
3. What is the median of the data in problem 1?
4. What is the mode of the data in problem 1?
5. What is the mean of the data in problem 1?
6. The data in problem 1 represent a sample drawn from a larger population. What is your best estimate of the population standard deviation?
7. In a certain city 10% of the residents are over 60 years old. What is the probability of finding no one over 60 in a sample of 10 residents?
8. Establish a 95% confidence interval for μ on the basis of the following sample statistics:

   $n = 100$
   $\bar{x} = 10$
   $s = 5$

9. You are working with samples of size 15 from populations that are approximately normally distributed. What table would you use to establish confidence intervals?
10. You suspect that two brands of soft drink differ in acidity. You measure the acidity of 10 randomly selected bottles of each brand with the following results:

Brand A	Brand B
$n = 10$	$n = 10$
$\bar{x} = 6.50$	$\bar{x} = 6.30$
$s = 0.05$	$s = 0.04$

    Outline an appropriate statistical test. What assumptions must you make, if any? Is the difference between the two brands significant at the 1% level?
11. "The power of this test against the alternative $\mu = 5$ is 0.80." Explain this statement.

12. The following data represent growth rates of plants before and after a chemical treatment. Outline an appropriate statistical test of the theory that the chemical treatment reduces growth rate. What assumptions must you make, if any? Are the results significant at the 5% level?

Plant	Before	After
1	1	3
2	6	3
3	4	5
4	5	7
5	4	0
6	2	1
7	2	5

13. Compute the correlation coefficient between the before and after scores in problem 12.

14. Use a regression equation with the data of problem 12 to predict the "after" growth rate of a plant with a "before" growth rate of 3.

15. On the basis of sample data you wish to determine whether the variances of two populations differ. What statistical test will you use? Write the formula.

16. You wish to test the theory that men have higher self-esteem than women. On the basis of a questionnaire you classify a group of men and women into categories of high, medium, and low self-esteem. The results are as follows: Perform an appropriate statistical test and comment on the results.

	High	Medium	Low
Men	10	30	10
Women	25	15	10

17. Complete the analysis of variance below. Are there significant differences between groups? ($\alpha = 0.01$)

	Sum of squares	df	Variance estimate	F
Total	580	29		
Between groups	175	2		
Within groups	405	27		

18. What assumptions are required for a two-way analysis of variance?

Answers

1.

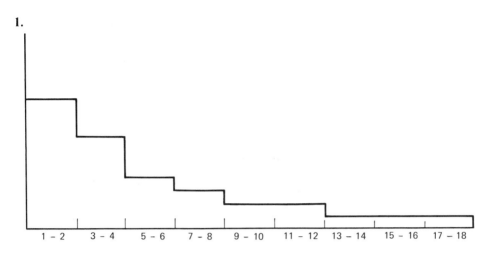

| 1 – 2 | 3 – 4 | 5 – 6 | 7 – 8 | 9 – 10 | 11 – 12 | 13 – 14 | 15 – 16 | 17 – 18 |

2. Skewed to the right

3. 4

4. 2

5. 5.6

6. 4.47

7. 0.349, or 35% of the time

8. μ is between 9.02 and 10.98 with 95% confidence.

9. "Critical values of t"

10. Null hypothesis $\mu_1 = \mu_2$
Alternative $\mu_1 \neq \mu_2$
Significance level $\alpha = 0.01$
Critical region $t \leq -2.78$ or $t \geq +2.878$

Assumptions: approximately normal distributions for both brands and approximately equal variances

$t = 9.901$; the difference is significant.

11. If $\mu = 5$, you have 80% probability of obtaining significant results.

12. Use difference scores
Null hypothesis $\mu = 0$
Alternative $\mu < 0$
Significance level $\alpha = 0.05$
Critical region $t \leq -1.943$

Assumptions: the population of difference scores is approximately normally distributed.

$t = 0$; the results are not significant.

13. $r = 0.22$

14. $b = 0.29$; the predicted "after" score is 3.31.

15. $F = s_1^2/s_2^2$

16. Use a X^2 test. $X^2 = 11.34$, which is significant at the 1% level. This result indicates that the distribution obtained was unlikely to have occurred by chance if there were no difference between men and women, but the results do not support the theory that men have higher self-esteem than women.

17.

	Variance estimate	F
Total		
Between groups	87.5	5.83
Within groups	15.0	

The differences between groups are significant.

18. Assumptions: all observations are independent; all groups have approximately normal distributions and approximately equal variances.

References

Question	Chapter	Question	Chapter	Question	Chapter
1	1	7	2	13	7
2	1	8	3	14	7
3	1	9	3	15	6
4	1	10	5	16	8
5	1	11	4	17	6
6	3	12	5	18	9

Index